# Industrial robot safety

HSG43

**HSE** BOOKS

First published 1988
Second edition 2000

ISBN 0 7176 1310 0



This guidance is issued by the Health and Safety Executive. Following the guidance is not compulsory and you are free to take other action. But if you do follow the guidance you will normally be doing enough to comply with the law. Health and safety inspectors seek to secure compliance with the law and may refer to this guidance as illustrating good practice.

# Contents

# Introduction

1 This book deals with the safeguarding of industrial robots. No new technical requirements have been introduced since the first edition was published in 1988. This new edition includes a section on relevant health and safety law and comments on the duties imposed by recent legislation.

## Robots and safety

2 The guidance includes advice for designers, manufacturers and suppliers on the principles of safe design for robot systems and the way in which they are likely to be used. The recommendations will be of use to users of industrial robots, whatever the size of their organisations, in helping them set up safe robot systems and safe working methods.

3 An agreed definition for a robot can be found in European Standard EN 775: 1992, as follows:

**'Manipulating industrial robot: An automatically controlled, re-programmable, multi-purpose, manipulative machine with several degrees of freedom, which may be either fixed in place or mobile for use in industrial automation applications'.**

4 For the purposes of this guidance, 'robot systems' means the robot and its control equipment as adapted to perform its particular function and includes any associated machinery and equipment. Although some robots are capable of locomotion, this is not generally regarded as being a defining characteristic. Such systems are outside the scope of this book.

5 This guidance is principally concerned with fixed industrial robots of the teach and play back type which use pendant controls for the teaching function. The range of uses for such robots is being widened but most are used for welding, painting, assembly tasks, machine loading/unloading and palletising. Some parts of this guidance are also relevant to other types of robot and to mechanical handling machines of the pick and place type.

6 Because of their power and commercial utility, there has been a rapid increase in the number of applications of robot systems in some industries. However, the very flexibility and complexity of robot systems introduce novel safeguarding problems. Most industrial robots have the capacity for inflicting serious injuries to anyone unfortunate enough to come into contact with them when they are working at normal speeds. It is relatively simple to identify likely hazards on robot systems. However, the way hazardous situations arise is not so obvious or predictable. It may therefore be difficult to select suitable safeguards.

## Standards and the law

7 These technical difficulties can lead on to problems in demonstrating compliance with current legal requirements relating to the supply and use of machinery, including robot systems. An integral feature of product safety supply legislation is the voluntary use of harmonised standards to assist in demonstrating compliance with essential health and safety requirements. Such harmonised standards are developed by European and international standardisation bodies, ie

- European Committee for Standardisation (CEN);
- European Committee for Electrotechnical Standardisation (CENELEC);
- European Telecommunications Standards Institute (ETSI);
- International Standards Organisation (ISO); and
- International Electrotechnical Commission (IEC).

8 This book recommends ways of safeguarding certain types of robot system which are compatible with advice provided in such standards. Where standards do not yet exist, the recommendations are based on sound engineering judgement and experience. A more detailed summary of legal duties is included in Appendix 1 and full details of legal publications, including relevant standards, are included in the References section.

9 The designer, manufacturer, supplier, integrator, installation contractor and user will have duties under one or more of the following provisions:

- The Health and Safety at Work etc Act 1974 (HSWA);[1]
- The Provision and Use of Work Equipment Regulations 1998 (PUWER);[2]
- The Supply of Machinery (Safety) (Amendment) Regulations 1994 (SMSR);[3]
- The Lifting Operations and Lifting Equipment Regulations 1998 (LOLER);[4]
- The Management of Health and Safety at Work Regulations 1999 as amended (MHSWR).[5]

Note: The risk assessment carried out for the purposes of MHSWR should identify the precautions necessary to comply with the relevant legislation.

10 All robots covered by this book are work equipment within the definition of PUWER and will be covered by PUWER requirements, eg concerning the stability, the adequacy of the controls and guarding arrangements. These

requirements will be met by following the guidance in this book. Regulation 6 of PUWER requires, for example, periodic inspection if deterioration is likely to lead to dangerous situations. Such inspections are particularly relevant to the stability of the robot and the integrity of the moving parts. Inspection of the safeguarding arrangements is also necessary to ensure that they continue to work correctly. Examples of these arrangements are the correct operation of interlocks and enclosure continuing to prevent access to a danger zone.

11 LOLER applies to robots which are used as lifting equipment, for example when carrying out lifting operations during mechanical handling and palletising. Paragraphs 28 and 29 of *Safe use of lifting equipment: Lifting Operation and Lifting Equipment Regulations 1998: Approved Code of Practice and Guidance*[4] make it clear that LOLER applies to a wide range of equipment and includes robots used for lifting operations. However, the introduction to the guidance stresses that the Regulations should be applied in a proportionate way that reflects the extent and nature of the risks involved.

**If the precautions adopted follow the guidance in this book, eg slow speed movement under load during teaching and enclosure of the robot during use, then the measures necessary to comply with LOLER are likely to be minimal.**

12 LOLER requires a risk assessment to be carried out to identify risks which may arise both during the lifting operation and at other times, for example during maintenance, teaching or cleaning of the surrounding area. LOLER also requires a thorough examination to be carried out in certain circumstances (eg if deterioration of the robot is liable to result in a dangerous situation) on those robots used as lifting equipment. The thorough examination should be carried out by a competent person and each case will need to be judged individually on its merits.

**For robots safeguarded in a way which follows the good practice described in this book, thorough examinations are unlikely to be necessary.**

# Principles of safeguarding robot systems

13 The type of robot, its use and its relationship to other plant and machinery will all influence the design and selection of safeguards. These will have to be suitable for the work being done and allow the operator, or operators, to carry out their work in support of normal operations safely. This includes feeding and removal of workpieces or components, loading magazines or feed devices, and dealing with interruptions in production, such as misfeeds or blockages. Where required, the safeguarding should also permit teaching, programming and setting. Many installations will require people to work in close proximity to the robot during such operations.

14 Other necessary work on robot systems includes inspection, servicing, maintenance, fault-finding and rectification. Where possible, such work should only be done when the dangerous equipment has been made safe by isolation. Where it becomes necessary to observe the robot closely while it is energised then additional safety measures, other than relying solely on the pendant control, may need to be taken. The chosen measures should be appropriate for the level of hazard presented by the installation. See paragraph 55 for guidance on safe systems of work.

15 Before designing or selecting particular safeguards, the hazards should be identified and the risk of injury assessed. For example, a small, low-powered robot not operating in conjunction with dangerous machinery will require a different level of safeguarding from a more powerful robot feeding machinery. To carry out an analysis it is necessary to:

- identify the hazards;
- evaluate the need for access and the risk to personnel;
- study possible failure modes and their influence on risk;
- identify a package of safety measures which eliminate or minimise the risks;
- decide on acceptable levels of safety integrity for the controls and safety devices; and
- assess the achieved levels of safety integrity for each part of the system and ensure that these achieved levels are acceptable.

The Glossary contains definitions of risk assessment and safety integrity.

16 Further information on hazard identification and risk assessment in the general context of robot systems is given in the section 'Identification of hazards and assessment of risk' (paragraphs 109-119). An example of their application is given in Case Study 3 in Appendix 2. Most types of safeguarding systems for industrial robots are based on a package of measures intended to:

- keep people at a safe distance from the robot whenever it is in operation or capable of movement; and
- only allow access when the robot and any ancillary equipment are in a safe state.

The case studies in Appendix 2 show some different ways of doing this.

17 Some operations may require close approach to the robot by certain people doing specific tasks when the robot is capable of moving, for example during teaching, program adjustment and observation of the work cycle.

18 Careful analysis should be made of the need for close approach to the robot. This should only be allowed and provided for in the design of the safeguards when there is no practical alternative. Further information can be found in the sections on programming and setting (paragraphs 68-80), and maintenance and troubleshooting (paragraphs 81-101).

19 When any close approach is necessary the integrity of the safety critical parts of the robot system should be of a very high order. This is particularly important when safety depends upon the correct functioning of the robot controller and its interfacing with the safeguards, irrespective of the technology these safeguards employ. The methods involved are covered in the section 'Interfacing with the robot controller' (paragraphs 161-199). It is not implied that only those methods described are suitable; other methods which provide equivalent safety may be used.

20 Where close approach to the robot is necessary during teaching, and for observation of the work cycle, the robot arm should be restricted to low speed (no more than 0.25 m/s - see paragraphs 154-156). The person teaching or observing should be in sole control of the robot.

21 At most installations it should not be necessary to approach close to the robot when it is operating at full speed as proper design of enclosures should allow observation of the robot or workpiece from the outside. Appendix 1 details the legal requirements to protect people in this situation and any proposal for close approach to the robot when it is operating at full speed would have to satisfy these requirements.

22 Some specific conventional safeguarding methods are listed in the 'Safeguarding methods' section (paragraphs 120-160). The combination of methods to be used in a particular application will depend on the assessment of the risk. It is unlikely that any single method will provide adequate safeguarding.

Information on risk assessment and the design of safeguards can be found in British, European and International Standards (see the References section). Seven case studies are given in Appendix 2; they show the variety of safeguards which can be employed to allow necessary access while ensuring safety.

## Safety in installation design

23 Safety should be a primary consideration at the planning and design stage of any robot system. This is when all those involved with the design, manufacture, installation and use of the robot system should exchange information and allocate responsibilities for ensuring safety.

24 Equipment and machine designers and suppliers should be familiar with the content of this guidance and any other published standards and guidance on safety matters appropriate to their product. Further information is provided in the 'Identification of hazards and assessment of risk' section (paragraphs 109-119).

25 New robots should be designed and constructed to comply with the requirements of SMSR as amended. Information should also be provided relating to the robot performance. The following features should be among those considered to ensure adequate safety of all robots:

- the design and construction of the robot should be appropriate for its duty and the expected environment, eg the effect of corrosive atmospheres and extremes of temperatures;
- components identified as safety critical should be of high reliability and sufficient safety integrity;
- all moving parts of the robot system should be designed or enclosed to eliminate or minimise possible trapping points;
- an adjustable speed facility on robot axis motions should be provided so that the end-user may teach and prove the program at reduced speed;
- brakes or hydraulic/pneumatic stop valves should be provided to prevent the robot arm moving when power is removed, eg after emergency stop;
- the control pendant should be ergonomically designed (see paragraphs 39-48 for advice);
- in addition to the defined limits programmed into the software, means should be provided to restrict the arcs of movement to the minimum required for the particular task by the use of fixed or adjustable positive stops - these may be physical stops so arranged that they do not present a trapping risk, or position detection switches operating into a hardwired safety system; and
- the forces exerted by the robot should be restricted to the minimum required for the particular task where the method of motive power allows this.

## Pneumatic or hydraulic power

26 Where the motive power is either pneumatic or hydraulic, the following additional design features should be included.

- pipe rupture valves should be fitted directly onto any supporting cylinders; and
- all fluid supply hoses should be enclosed so that they are protected from mechanical and/or environmental damage.

## The robot control system

27 For the control system, safe design should prevent any unforeseen movements which could create a hazardous situation. The control system should be designed so that following any interruption of the program, movement of the robot should only restart under known conditions. The method of returning to normal operation will differ according to the nature of the interruption and may include:

- giving a known command, eg at the end of the cycle;
- following a set procedure, eg after emergency stop; or
- by manual commands to a parked or home position, eg after a cycle interrupt or emergency stop.

28 It should not be possible for the robot to move to the parked or home position under program control by an unspecified trajectory. Any detected unexpected movement, lack of movement on demand or movement faster than the pre-set limit should remove all power from the robot axes and a fault condition should be indicated.

## Design of grippers

29 Gripper systems should be designed so that:

- power failure, including loss of hydraulic or pneumatic pressure or activation of the emergency stop circuit, does not cause the gripper to release the load;
- the gripper is designed for the maximum dynamic force of the load, including that imposed by emergency stopping;
- the static and dynamic forces created by the load and gripper together are within the load capacity and dynamic response of the robot; and
- any increase in the extent of the robot operating envelope and thus of the danger zone, caused by the gripper, is clearly established and made known to those responsible for installing the robot.

# Safety at the design stage

30 In many cases more than one supplier of equipment may be involved. Responsibility for safety of the complete installation should ideally be given to a single person. This will often be the turnkey supplier who has the responsibility for purchasing, installing and commissioning the complete installation, although the user may sometimes elect to install the equipment themselves or may rely on the robot manufacturer or supplier.

## Installation layout

31 Care in planning the layout can eliminate many hazards. The aim should be to minimise the need to approach danger zones by providing good arrangements for viewing from outside the enclosure and by providing means for feeding and removal of components so that no-one need enter the enclosure. (See Case Studies 1, 2, 3, 4 and 7 in Appendix 2 for examples of how this may be done.) Safeguarding arrangements as exemplified in these case studies are normally sufficient to discharge the duty, required under LOLER regulation 6, to reduce the risk of people being struck by the robot or any load it is carrying.

32 The layout of the enclosure and the safety arrangements should allow for access to an injured person who may need to be assisted or removed.

33 Access to the enclosure should only be possible through interlocked access gates, or their equivalent, which will stop all dangerous movement of the robot and of associated equipment unless the associated equipment is effectively safeguarded or presents no danger.

34 When access to the enclosure is required with the robot under power and capable of movement, eg for programming or maintenance, the design should allow adequate space around the robot where personnel can stand safely. It is important to remember that an area outside the programmed reach of a robot may not be safe in the event of malfunction. Attention should be paid to the potential hazards of all the associated equipment - this applies to both direct hazards and those which may result from abnormal behaviour. Local fixed or interlocked guards may be necessary to eliminate the hazard.

## Design for maintenance and fault-finding

35 Safety for maintenance staff should be a criterion of the design. Whenever possible, designers should arrange for maintenance to be carried out with the power 'OFF' and with a minimum disruption to in-built safety features. When entrance to hazardous areas is required with power 'ON' to the robot, appropriate lockout and interlocking arrangements should be maintained between the motive power circuits and those components causing dangerous movement. This arrangement is not always possible and it may be necessary to carry out maintenance with the controls set at power on/servo-hold. Servo-hold is the stationary condition achieved and sustained through the robot motion controller, rather than relying on hardwired interlocking to prevent robot movement. In these circumstances additional safety measures will be required, eg monitoring of the servo-hold condition (see paragraphs 193-195).

36 User-friendly diagnostic facilities such as error codes on VDU screens or indicator lamps on a diagnostic board will enable maintenance staff to identify fault areas. The fault monitoring system may be able to automatically override the operational program and place the robot/machine functions on servo-hold until appropriate action is taken.

## Controls

37 Control panels, keyboards and terminals should be located outside the enclosure and in such a position that the operator has a good overall view of the work process. In some installations this may require an elevated platform, or monitoring by closed-circuit television. Where possible, loading and unloading points should be near control panels and terminals. It should be possible to control machinery or plant associated with the robot system from the main control console. If this is not possible, additional emergency controls should be provided.

38 Controls should be identified and labelled in accordance with BS EN 60204-1, where appropriate, and should be ergonomically laid out to minimise errors. Colour should not be used as the sole or primary method of identification, but is effective when combined with other identification methods. Red should be reserved for emergency meanings and emergency stops should be provided at each workstation.

# Teach pendant design - ergonomic aspects

39 The application of ergonomic principles to the design of teach pendants can improve safety by simplifying tasks and reducing the scope for human errors. Some of the relevant points are given below.

## Pendant - dimensions

40 If pendants are to be held for long periods, they should be as small and light as possible. The centre of gravity of the pendant should act through a point held by the hand and the area of the pendant where most keying takes place should be placed near the centre of gravity.

## Connecting cable

41 The type of cable is important since if it is too light it may easily be damaged, yet if too heavy the pendant may feel heavy or unbalanced. It should not enter the pendant at a point which causes twisting and makes it uncomfortable to hold. A cable dispenser will help keep the cable tidy and away from places where it may be damaged.

## Number of controls

42 Controls on the teach pendant should be kept to a minimum. Some can be placed on a main control panel without causing inconvenience to the operator. An emergency stop should always be provided on the pendant.

## Multi-function controls

43 Multi-function controls are confusing and slow to use and should be avoided on a teach pendant where possible, especially on controls causing robot motion.

## Control action

44 Controls which cause robot movement should be of the hold-to-run type which, when released, cause movement to stop. Controls should have a positive action so that when switching occurs it can be felt or sensed.

45 Labels should be located systematically in relation to controls and should be brief; only very common abbreviations should be used. The control label should be visible when the operator is using the control. Symbols should be clear, permanent, simple and commonly used.

## Control layout

46 If the controls are normally operated in a particular sequence they should be placed in a logical order. It is usually better to do this than to group them functionally.

47 To prevent accidental actuation, controls should be adequately spaced and important controls can be recessed, covered by a latch, locked or interlocked.

48 Emergency stop controls should comply with BS EN 418. On pendants, emergency stop actuators should be of the mushroom-headed type.

## Operator position

49 The operator should know where to stand in relation to the robot to obtain the correct control orientation. Robot left is not always operator left. Aids to orientation, including floor marking, can assist in correct use of the teach pendant.

# Providing information

50 Providing adequate information on the safe installation, use and maintenance of the robot and/or robot system is required by SMSR and will be part of the designers', manufacturers' and suppliers' responsibilities. Similarly, PUWER 1998 places duties on employers to provide information to enable people at work to use and maintain the robot safely. Information should include:

- a clear, comprehensive description of the robot and its installation;
- guidance on the principles of operation, the frequency of inspection, frequency and method of functional testing and guidance on the repair and maintenance of safety devices;
- guidance on fault-finding and rectification including safety procedures;
- maintenance schedules including safety procedures;
- clear guidance for the safe programming of the robot including use of the teach pendant;
- software documentation; and
- to aid maintenance and repair, a program for full articulation at a suitable low speed and within the limits of the particular installation of each motion of the robot.

51 It is strongly recommended that the supplier of robot systems provide comprehensive training information to the user and gives assistance in the training of the user's personnel.

# How to install, commission and test safely

52 During installation and commissioning two main safety needs arise. They are:

- establishing safe working conditions and procedures during a period of change; and
- reviewing the long-term safeguards for the completed installation as provided for in the design. If modifications to the safeguards are required, then they should be subject to the same verification and validation procedures as were used in the original design phase and fully documented.

## Installation

53 The installation stage presents changing hazards as the robot system is built up. For example, robots are inherently unstable and care is needed when positioning until they are bolted down. Safety requirements should be assessed and reassessed frequently as installation proceeds. This may best be achieved by identifying who has overall responsibility for the project, including workplace safety. Frequently, this will be the turnkey supplier (see paragraph 30).

54 The project leader should ensure that:

- chains of responsibility are clearly set out and understood by all concerned (this may include systems of work and/or written authorisation procedures);
- common safety rules are established for the various contractors present;
- the special needs of contractor staff and others not normally employed at the permanent site are provided for;
- the safety of all staff is ensured during the period before the full and final safeguards are installed, eg by providing additional temporary safeguards; and
- in the case of a robot used as lifting equipment, an assessment is made of the need for a thorough examination as required by LOLER regulation 9(2) before the robot is put into service.

## Commissioning

55 Both the installer and the user should be satisfied that the robot is installed and working properly before it is handed over for production. A safe system of work will be necessary to ensure that commissioning personnel can carry out their tasks in a safe, controlled manner and to meet their obligations under PUWER regulation 7. A safe system of work is usually specific to an installation but will include:

- a written plan of work which is known to all involved and which lists tasks and divisions of responsibilities;
- written procedures understood by all concerned for each of the tasks listed in the plan;
- nomination of authorised staff for isolating and securing all types of power supplies, for entry to enclosures/working zones and for use of override manual controls;
- selection of competent staff and provision of proper training in all aspects of maintenance work including safety training;
- a procedure for calling in the supplier's field experts where serious or complicated breakdowns occur;
- proper documentation including:

  - a description of all procedures defined under the system of work;
  - records of corrective work carried out;
  - a system for recording and analysing faults to enable appropriate action to be taken;
  - record of any changes in design or operational procedures implemented;
  - record of modifications to the plant or its control equipment; and
  - amendment of procedures where necessary to take account of changes made.

## Testing

56 The supplier should provide a comprehensive schedule against which the performance of the safety-related functions of the robot installation should be tested.

57 Faults discovered while testing safety functions should be recorded, investigated and rectified before the robot system is taken into use. A robot system should not be taken into use with a system of work substituting for a faulty safety function.

58 Records of all performance tests, including the conditions under which the tests were made, should be kept to form part of the supplier's handover documentation.

# Handover documentation

59 The following documents should form part of the handover package:

- overall specification of the robot system;
- operational instructions;
- routine maintenance instructions;
- list of recommended spares;
- list of recommended specialist tools/diagnostic aids;
- fault-finding procedures, including use of any self-diagnostic capability;
- records of any modifications and/or changes made during commissioning;
- records of acceptance tests; and
- the EC Declaration of Conformity or a Declaration of Incorporation.

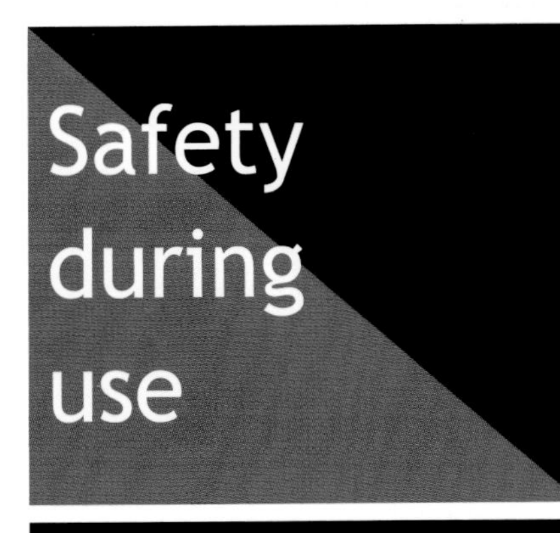


60 Robots are designed to be strong and stable enough to carry out the required operations when used in accordance with their specification. In particular, where a robot is used as lifting equipment, LOLER regulation 4 requires that its strength and stability are adequate during use.

## Normal production

61 During normal production, the system will often operate unattended and therefore close approach should be unnecessary. If the hazard identification and risk assessment have identified hazardous situations that may arise from the use of the robot, the safety measures should prevent operators, material loaders, cleaners, labourers and passers-by from gaining access to the installation and being exposed to potential injury.

62 During normal operation with the robot in automatic mode, the operator may be required to approach the robot and associated tooling. This should only be possible when the robot and associated machinery are in a safe state. In these circumstances, a safe system of work may also be required, the extent and detail of which will depend on the activity to be carried out. It is preferable that the access arrangements require:

- a controlled shutdown through the robot controller as this makes re-start easier, and
- power to the robot axes to be cut off through a means of isolation independent of the robot controller.

63 If power has to remain 'ON', the interlocking arrangements should ensure the robot remains stationary. See the 'Safeguarding methods' section (paragraphs 120-160) for descriptions of safeguarding methods and interlocking, and the case studies in Appendix 2, particularly 2, 3, 4, 5 and 7, for descriptions of access arrangements.

64 If an operator is required to work on a single robot while other adjacent robots continue to operate, safeguarding arrangements should be provided to prevent the operator gaining access to, or being injured by, those other robots. Fixed barriers or electro-sensitive sensing devices may be required. The safety implications of material being handled which protrude beyond this safeguarded area should form part of the overall risk assessment.

# Safety at associated machines and equipment

65 The risk posed by associated machines and equipment, including feed, transfer and delivery machinery, should be assessed and, where appropriate, safeguards applied, taking into account any relevant standards. This is particularly important if associated machines and equipment can be run independently of the robot, when access is gained to the installation. (See Case Study 3 in Appendix 2.)

66 Safe working procedures for setting associated machines should be developed, using information from the suppliers of the machine or installation. When there is a risk that setting associated machines and equipment could initiate movement of the robot, the robot should be isolated from its power source and the interlocking arrangements should not be relied upon for isolation.

67 In large installations where a robot is working with several machines and the remaining machines are required to operate while one is being set or serviced, special safety precautions should be established. This may involve the use of alternative sub-routine programs and the provision of temporary or permanent safeguards, eg barriers. Safe systems of work should be established to ensure that isolating procedures are followed.

# Safety during programming

68 The embedded software provided by the robot or installation manufacturer or supplier should not be accessible to unauthorised people. It should be protected against corruption during applications programming. Applications programs should be written and proved by people competent to do so, and access to them restricted to authorised people.

69 Program-proving procedures should be developed using, for example, computer graphics, single-step facilities, and slow-speed and full-speed dry runs (ie without component or material). Records and documentation of changes to the program should be maintained as part of a formal change-control procedure.

70 Programming involves the teaching or correcting of position. Such teaching is often done from a pendant control. The design of the system should be such that all or most of the teaching can be done from outside the enclosure, eg by ensuring good visibility. If it is necessary to do some teaching, or to observe part of the robot cycle from within the enclosure, the following measures should be taken:

- there should be a positive means of switching from automatic operation to teach mode and the means of restoring full power for automatic mode should be held by the person doing the programming in the cell. This system should preferably be achieved automatically by design measures. See Schedule 3, Requirement 1.2.5 of SMSR. Methods may include:
  - a trapped key exchange system (see Case Study 3 in Appendix 2 for a description of one method);
  - automatic transfer of control by key switch or password from the console to the teach pendant and from automatic to teach mode when the pendant is activated;
  - switching by key from automatic mode to teach mode, in which the operator retains the key and where access is not possible to the robot enclosure with the robot on automatic mode;
- if reliance is placed on keeping interlocked access gates open as part of the safety procedures to prevent the robot being capable of operation in automatic mode, they should be prevented from closing, eg by use of a trapped key exchange system, or by other physical means. Warning signs should be used to deter others from entering during teaching;
- when the teach pendant has been activated it should only be possible to control the robot from the teach pendant itself. The controls on the main control panel and any external signals should have no effect. As an exception to this, all emergency stop controls should always remain active;
- clear working space of 500 mm minimum should be provided around the robot operating envelope and around associated machinery and equipment to enable teaching to be done from a safe position. See BS EN 349 on prevention of crushing of the human body;
- speed should automatically be restricted and, where possible, low torque/force selected, unless normal operating speeds are slow. What constitutes an acceptable slow speed will depend on the layout of the installation and forces exerted by the robot, but a speed of more than 0.25 m/s is unlikely to be acceptable, although teaching is often done at speeds far slower;
- the following should be provided:
  - hold-to-run teach controls for initiating robot movement; and
  - an independent emergency stop.

71 Teach pendants should incorporate an enabling device (see paragraph 152 for description) which permits the hold-to-run controls to initiate movement when it is held in a particular position. Release of the device should arrest all motion. During programming it may also be necessary to synchronise the operation of associated equipment and, where necessary to prevent danger, this equipment should also be put in a safe state before entering the enclosure.

# Program verification

72 Verification of the program which has been taught is usually necessary after teaching or reprogramming or as an in-service check of production accuracy.

73 If visual examination of the complete program cycle is a necessary part of the verification procedure it should, whenever practicable, be made from outside the enclosure.

74 If it is necessary to observe the program cycle from within the enclosure, robot motion should be at reduced speed and all the precautions for safety during teaching listed in paragraphs 70 and 71 should apply.

75 There may be some exceptional cases where it is necessary to observe robot movement at full operational speed from within the safeguarded area. This will depend on the application and the specific task and arrangements for doing this work will have to comply with PUWER, in particular regulation 7, and HSWA section 2 (see Appendix 1).

76 Close observations at full speed should only be done after a full assessment has been made by a responsible person, who has considered whether there are alternative methods which might be used.

77 Those tasks where close observation at operational speed is necessary should be identified and safe working procedures drawn up. These procedures should ensure that approach is restricted to the minimum time necessary to ensure verification and should identify a safe place or places from which this work can be done.

78 Teach pendants should not have the facility to operate the robot at full speed for work within the enclosures unless it has been specified as a result of the assessment by a responsible person.

79 In such cases it will be necessary to override the normal teach function. This may be achieved by using an additional feature on the teach pendant or by a separate control device. In either case the following requirements apply:

- reduced speed should only be suspended by means which require a deliberate and intentional action by the teacher;
- there should be an enabling device. This may be part of the teach pendant or may be a separate device; and
- the enabling device should be hardwired.

See the 'Interfacing with the robot controller section' (paragraphs 161-199) for details of how to achieve an appropriate safety integrity level for teaching.

80 On completion of verification, reinstatement of normal operational control should not be possible until the full safeguards have been restored.

# Safety during maintenance

81 Maintenance covers the following activities:

- routine inspection and servicing;
- planned preventive maintenance;
- fault-finding (sometimes known as troubleshooting);
- rectification; and
- major repair or modification.

There are also the statutory requirements of LOLER and PUWER which may apply, depending upon individual circumstances (see paragraphs 10-12).

82 In some robot installations, planned preventive maintenance, fault-finding and rectification is provided by the robot manufacturers or specialist contractors. In larger organisations, in-house staff might be used for the routine inspection and servicing role. The division of responsibility for the above operations should be clearly defined and documented by the user.

83 Safety during maintenance is important because close approach to the robot is likely to be necessary. The main safeguarding objective should be to allow staff to gain access for legitimate maintenance activities, while preventing operation of the robot system in the production mode.

84 Safe access for servicing should be a design consideration (see paragraphs 35 and 36 on design for maintenance and fault-finding). The complex nature of robot systems requires that users establish well thought-out procedures which protect maintenance staff. This should be based on information provided by the supplier.

**The maintenance of robot installations should be done by those who have received adequate information, instruction and training relating to that work. Maintenance workers should have adequate levels of skill, work in a logical sequence and, whenever possible, work with the power supplies to the robot system locked off.**

85 Employers should provide written instructions for those involved in the maintenance of robot systems. Safe systems of work should be drawn up for all maintenance work. These should include the nomination of authorised staff for isolating and securing power supplies. A permit-to-work system may be required. Restoration of power should include a check of safety devices.

86 Maintaining robot installations can be particularly dangerous because of the complexity of robot functions and the possibility of unpredictable events. These include:

- unpredictable movements while being maintained or at the reinstatement of power;
- corruption of software, or damage to the programmable electronics and peripherals during maintenance;
- problems arising from interfacing with other machinery and equipment. There will often be a need to isolate part of a complex robot system to sustain productive capacity on related machines. Hazards may arise when there are interfering trajectories of adjacent robots or equipment, or when taking associated machines and equipment out of programmed sequence;
- the need to provide known conditions on start-up after maintenance; and
- hazards arising from the functional purpose of the installation (eg the robot arm may be carrying an unguarded abrasive wheel or high-pressure water jetting equipment, or a laser).

87 During maintenance, care should be taken to ensure that there is no danger arising from movements of any other robot system or associated machines and equipment still in operation. Where this is done by changing to an alternative control program (application program), a proper system of working should be established between the production supervisor and maintenance personnel **before** work begins.

88 On large complex installations where one or more robots may be serving a number of stations, access for servicing may be required to one part of the installation while remaining sections continue to operate normally. Servicing workers should be fully protected from the rest of the working robots or machinery. In this situation the working robots should be prevented from approaching the area where servicing work is taking place and there should be no automatic transfer between the working and non-working areas. Temporary barriers or other equally effective measures should be provided. The possibility of component ejection should also be considered when designing barriers.

89 All equipment should be returned to a safe position before power is restored. Known starting conditions should be provided for start-up. Frequently, robot systems carry powered tools on their manipulating arms. Maintenance

personnel should be aware, before carrying out any maintenance work, of any dangers arising from inadvertent operation of such tools and should take suitable precautions.

90 Local emergency stops and means of isolation should be available for each part of an integrated plant where sectionalised access is likely to be required. Maintenance staff should take care not to interfere with the operation of sensors which are working in adjacent sections of the installation.

91 Adequate information and manuals on the robot system and any related equipment should be provided by the supplier. Manuals should contain adequate design information including procedures for carrying out routine inspection and servicing, planned preventive maintenance, fault-finding and rectification.

92 All maintenance staff should be provided with appropriate personal protective clothing and equipment for their tasks. Suitable tools, mechanical aids and lifting tackle should also be provided as appropriate.

93 Provision for planned preventive maintenance should be made at the design stage and should aim to minimise the risks to staff during those activities.

94 When working on the robot, the power supplies to the robot system should be locked off. In addition to any interlocking arrangements, it is necessary to isolate and lock out the actuating power supplies. Residual power should be dissipated and the robot should be placed in its lowest physical resting position (or alternatively provided with chocks).

95 Following any maintenance work which disturbs safety devices or guards, an independent check of their operation should be made before production is resumed. The condition of safety devices such as gate interlocks will deteriorate with use. All safety devices and safeguards must be regularly inspected and their effectiveness checked in accordance with regulation 6(2) of PUWER. All maintenance personnel should be in a safe position before the robot installation is restarted.

96 Safe working during fault-finding and emergencies should be provided for at the design stage. This is particularly important for complex installations where robot systems are used in combination with dangerous machinery and plant.

97 Maintenance staff are unlikely to be present when a fault occurs. They should seek to establish the history of the fault and the point in the program when it occurred. Any attempt to operate the robot to get a clearer picture of the fault should be carried out from a safe position with all safety devices operative. Full use should be made of any diagnostic facilities provided.

98 Entry into the robot working space as a result of an unforeseen stoppage is potentially dangerous should the fault clear and allow the cycle to continue. Therefore, close approach should only be made following normal access procedures through interlocked gates or openings protected by safety devices which stop the cycle and prevent it from resuming until the person has left the enclosure and safeguards reinstated.

99 There may be situations where personnel require access to the working area of the robot and associated machinery and equipment with power 'ON', for example to closely observe robot or machine functions. A safe system of work will be required for such access. With large multi-robot/machine installations a formalised permit-to-work system may be necessary. In all installations, entry under such conditions should only be made after the engagement of controls which permit only manual operation of limited (or slow) movements of the robot and any associated machinery.

# Safety during modification

100 Modification to the robot system or its control equipment should be properly recorded. The impact of a modification on other safety-related systems is often overlooked. Where modification is contemplated, change-control procedures are needed. Such procedures may require a documented hazard and risk analysis to be undertaken before the change is authorised. The design specification for the modification should then be subject to a design review procedure at least as rigorous as that applied to the original design.

# Documentation

101 The actions taken as a result of all routine inspection, servicing, maintenance and modification should be recorded. If an inspection has to be carried out for the purposes of PUWER regulation 6 (see paragraph 10) then it should be recorded as required by PUWER regulation 6(3). Any reports of thorough examination carried out under LOLER should be kept available for inspection.

# Training

102 Mistakes caused by lack of or inadequate training can lead to accidents. Safety training should cover:

- the basic procedures necessary to operate the robot system safely;
- the hazards entailed in using the equipment; and
- precautions to be taken.

103 Those who program, teach, operate, maintain, rectify and supervise or manage a robot system should be adequately trained for the job they are expected to do, and the precautions to be taken.

104 Other people, such as cleaners and contractors, who work in the vicinity of robot systems should be given training appropriate to their needs.

105 In considering the training required, take account of the circumstances in which people have to work, eg alone, under close supervision of a competent person, in a supervisory capacity etc. Training should include:

- a clear definition and description of the work;
- identification and explanation of controls likely to be used;
- identification of the hazards associated with the installation in general, and with the work to be done;
- a description of the safeguarding methods, and how they should work; and
- a description of any safe working procedures in force.

106 Hands-on training should be a controlled activity, under the effective supervision of a person who fully understands the system and the task to be performed. Onlooking trainees should be outside the robot system enclosure, or effectively protected by safety devices which allow observation but prevent approach to dangerous areas. (See Case Study 5 in Appendix 2 for a description of a robot system in an education establishment.)

107 Suppliers and users should work together to draw up an initial training programme but it is the duty of the user to ensure that adequate training is provided.

108 Training is an ongoing requirement. Operational and safety training should be reinforced by refresher courses and by regular reminders of safe routines in instruction documents. If routines are changed, then further training should be given.

## Risk assessment

109 Risk assessment is a series of logical steps which enables a systematic examination of hazards and the likelihood of their occurring. This process allows a judgement to be made about the level of protection required.

110 Robots are different from conventional machinery in that the working envelope is not readily visible and it is therefore difficult to anticipate its movements and the full extent of the danger area. Furthermore, even when they are operating according to program, their sequence of movements, stops and starts are not easy to predict.

111 Personnel carrying out functions such as setting, programming and servicing may be particularly at risk from robot systems. The first step in ensuring that these people are adequately protected is to identify all the hazards present in the robot installation. This should be followed by an assessment of the risk from each hazard. Depending on what this shows, risk reduction measures may be necessary.

**HSE recommends a five step approach to risk assessment:**

**Step 1:**

Look for the hazards

**Step 2:**

Decide who might be harmed and how

**Step 3:**

Evaluate the possible degree of harm arising from the hazards and decide whether existing precautions are adequate, or more should be done

**Step 4:**

Record your findings

**Step 5:**

Review your risk assessment immediately it becomes known that a change has been made which may make the original risk assessment invalid. In any event, you should review your risk assessment regularly and revise it if necessary

For more information, see HSE leaflet *Five steps to risk assessment*.[6]

## Hazard identification

112 Hazardous situations and events are identified by a detailed and systematic scrutiny of the robot installation. These will include those hazards associated with all phases of use, including assembly, dismantling, fault conditions and foreseeable abnormal situations. (Annex A of BS EN 1050 gives a detailed listing of hazards.)

113 It is necessary to consider abnormal behaviour especially, but not exclusively, of the robot itself. Abnormal behaviour can be defined as any unintentional or uncovenanted movement caused by a malfunction of the control system, or programming error.

114 Because of the complexity of most robot systems, hazard identification and risk assessment should be considered jointly by users and robot suppliers and should be carried out at an early stage of the project. Similarly, managers of turnkey projects should ensure that the hazards in the total system are assessed at the beginning of the design phase. Users of existing robot installations also have a duty to carry out and review their risk assessments to ensure continued safe operation.

115 Hazard identification should include:

- a study of the overall operation of the robot, including any working practices and its associated machinery, equipment and processes if any. This should cover the normal running of the installation and all other foreseeable circumstances during teaching/programming, setting, tool changing or adjustment, servicing or other reasons necessitating close approach;
- consideration of the size, power and operating characteristics of the robot and any particular hazards arising from them (eg pneumatic equipment may have an air reservoir and could be subjected to sudden movements of one or more actuators when a jam-up is freed, or hydraulic equipment may present a fire hazard if fluid is released);
- consideration of the functions of the robot which can affect safety during normal operation

and for any foreseeable fault conditions (these could include control valve malfunction or electrical faults);
- an examination of each function of any associated machinery, plant or process, each interaction with the robot and any interaction with personnel;
- the nature and degree of severity of each foreseeable hazard; examples are trapping, striking or entanglement etc. The robot end effector, with its gripper, may present other hazards. Any process hazards (toxic fumes, welding flash, lasers, dust, coolant, high-pressure water jet etc) should be taken into account; and
- the effect of hardware and software failures in the programmable electronic systems being used to control each element of the installation as well as any which may be supervising the operation of the entire installation. This should include the potential effect of electromagnetic emissions on the programmable electronics from, for example, the use of mobile phones.

# Risk estimation

116 Once all hazards have been identified, an estimate of the risk should be made for each hazard. This can be done by considering the following elements:

- the severity of harm; and
- the likelihood of that harm occurring, which depends upon
  - how often a person is exposed to to the hazard;
  - the probability of a hazardous event occurring; and
  - the possibility of avoiding or limiting the harm.

When estimating the risk, both technical and human factors need to be considered.

# Risk evaluation and reduction

117 Once the risk has been estimated, it has to be evaluated to determine how much it needs to be reduced to achieve safety. The best solution is to eliminate hazards through changes in design or layout (intrinsically safe design). If this cannot be done, then safeguards should be provided. These safeguards may include fixed or interlocked guarding, electro-sensitive protective equipment, or presence-sensing devices and can be augmented by other means including speed limitation and two-hand control devices. Any remaining risks will have to be dealt with by systems of work, training and information. The supplier should provide information about these to the user who then has the duty to carry them out.

118 Industrial robots may also be displayed at exhibitions or used in educational or research areas. The general public and students may therefore be at risk. Safeguards need to be chosen which allow viewing while ensuring protection. At exhibitions and for display purposes, routines may be devised which avoid traps between robots and adjacent equipment and which do not present a risk of ejection of components. Fixed transparent barriers may be adequate.

119 Case Study 5 in Appendix 2 shows an installation suitable for education or research where, in addition to viewing, there may be a need for adjustment or measurement, and where more sophisticated and flexible safeguards than those for public exhibitions may be needed.

# Safeguarding methods

120 This section describes conventional safeguarding measures for robot installations. Applications which pose a risk to those outside the enclosure such as laser or water jet cutting or emission of dust and fume, will require customised solutions. The safeguarding measure or combination of measures chosen will depend on the use to which the robot is put, the type of hazard it creates and the level of risk. The measures that may be taken have been put into a hierarchy of four levels:

- fixed enclosing guards;
- other guards or protection devices;
- protection appliances (jigs, holders, push sticks etc); and
- the provision of information, instruction, training and supervision.

In selecting safeguarding measures, it is necessary to consider each in turn from the top level, and use measures from that level as far as it is practicable to do so.

121 BS EN 292 gives information on risk assessment, safeguarding methods and how to select an appropriate safeguard for that level of risk. This and associated standards should be consulted when deciding which of the following safeguards are suitable for a particular application. Factors such as the stopping speed and distance and any effect of gravity on the robot need to be taken into account when designing the safeguarding equipment. There is additional detail on how safety devices can be interfaced with the robot controller in the following section (paragraphs 161-199).

122 The safeguards described below and in the following section are not the only ones which may be used in any particular situation. It should, however, be possible to demonstrate that any alternative methods provide a similar level of protection to those illustrated in this guidance.

## Perimeter fencing

123 Conventionally, perimeter fencing is made from 2 m-high, rigid panels securely fastened to the floor or to some convenient structure and positioned so that it is not possible to reach any dangerous parts of machinery. It should only be possible to remove fixed fencing with the aid of a tool.

124 The fencing and its fixing should be strong enough for the particular need. For most purposes, a hollow section steel framework in-filled with mesh is satisfactory. Where there is danger, for example from molten metal splashing, welding flash or ejected components, the in-filling should be made from sheet panels, reinforced if necessary. A combination of materials may provide a good solution (eg tinted plastic curtains and steel mesh at welding robots).

125 Where regular access is required, the fencing may be provided with sliding or hinged interlocked access gates; and/or an opening protected by a trip device.

126 Where it is foreseeable that the normal access opening is either not large enough or is inconveniently placed for maintenance work, sections of the fencing may be made demountable, providing they can be removed only with the aid of a tool.

127 Temporary fencing may be used where it is required to provide a short-term safe area within a guarded enclosure (eg to provide a safe area around a malfunctioning robot in a vehicle body welding line while continuing production). The temporary fencing should:

- be of suitable height and prevent access to continuing dangerous movements (see BS EN 294);
- be of stable and robust construction (see BS EN 953);
- be interlocked with the appropriate part of the safeguarding system; and
- be used in conjunction with a safe system of work.

## Interlocking devices

128 Examples of interlocking devices for access gates are:

- guard-operated interlocking devices designed for use in safety systems, eg positive and negative mode cam-operated, position-detecting switches which are cross-monitored and operating in the control circuit of the robot;

- key-interlocking devices, trapped key-interlocking being a common method. This is an interlocking device relying upon the transfer of keys between a control element and a lock fixed on a guard. These are mainly used for removing the power supply to all or parts of the robot installation. There are a number of ways this can be done but the essential elements are two locks, the first on the gate(s) in the perimeter fencing and the second on the robot controller. In the case of hydraulic and pneumatic equipment, the second lock would need to be on a valve actuator. The key cannot be removed from the robot controller (or valve actuator) lock to open the gate lock until a safe condition is established (see the following section 'Interfacing with the robot controller');

- a solenoid lock - this is an electromechanical device activated by an electrical signal. When the signal is received the locking bolt is withdrawn, allowing the access gate to be opened or a second key to be released which in turn releases the gate. The device may also be used in conjunction with a time delay. (See the 'Interfacing with the robot controller' section.)

129 On complex installations with perhaps more than one robot and other associated machines and equipment, the whole system may involve more than one access gate and isolation point and may require many keys and locks. (Further information on interlocking devices is given in BS EN 1088.)

# Electro-sensitive safety systems

130 Modern electro-sensitive protective equipment (ESPE), particularly when used in conjunction with conventional fencing, can provide a high standard of safety while allowing the necessary access. ESPE may also be suitable for use in situations where conventional guards are not practicable. The equipment may operate as trip devices detecting the approach of people or objects to the danger zone, or as presence-sensing devices where the dangerous operations cannot take place so long as a person or object is detected.

131 Safety depends on the electrical integrity of the ESPE, its location with respect to danger zones and the electrical and mechanical integrity of the rest of the system. In high-risk situations, certain machine functions, such as the stopping performance and the performance of the devices controlling the dangerous motions, may have to be monitored. If this function is not part of the ESPE, then a separate monitoring device may be required. BS EN 61496 Part 1 is a specification for the design and testing of ESPE.

# Light curtains and light-beam devices

132 These are the most commonly used sensing devices. They operate by detecting an obstruction in the path taken by a beam or beams of light. The intangible barrier created by this system may consist of a single beam or a number of beams of light, a curtain of light, or any combination of these. The light may be visible or invisible, eg infra-red, and may be continuous or modulated, eg a scanning system. European standards refer to these devices as active opto-electronic protective devices (AOPDs).

133 Recommendations for the design and performance of high-integrity light curtains are given in IEC 61496 Part 2. Guidance on positioning and use of light curtains is given in BS EN 999 and the HSE publication *Application of electro-sensitive protective equipment using light curtains and light beam devices to machinery.*[7]

134 Light curtains have three basic uses in safeguarding systems:

- as a trip device where the light curtains are usually arranged in a vertical format;
- as a presence-sensing device where the light curtain is usually arranged in a horizontal format; and
- a combination or zoning system where two or more may be used as sensors and/or trip devices to provide a more complete or selective protection against access to the moving robot or associated dangerous machinery.

135 Figures 1 and 2 show two applications. In Figure 1, two vertical light curtains are used selectively to permit access to the pallet while the robot is attending the conveyor or is parked in Zone Y. During production, light curtain A is activated and light curtain B is deactivated (by the controller) to allow the robot to enter Zone X. When the robot is in Zone Y, light curtain A can be deactivated manually so that a pallet can be moved into Zone X.

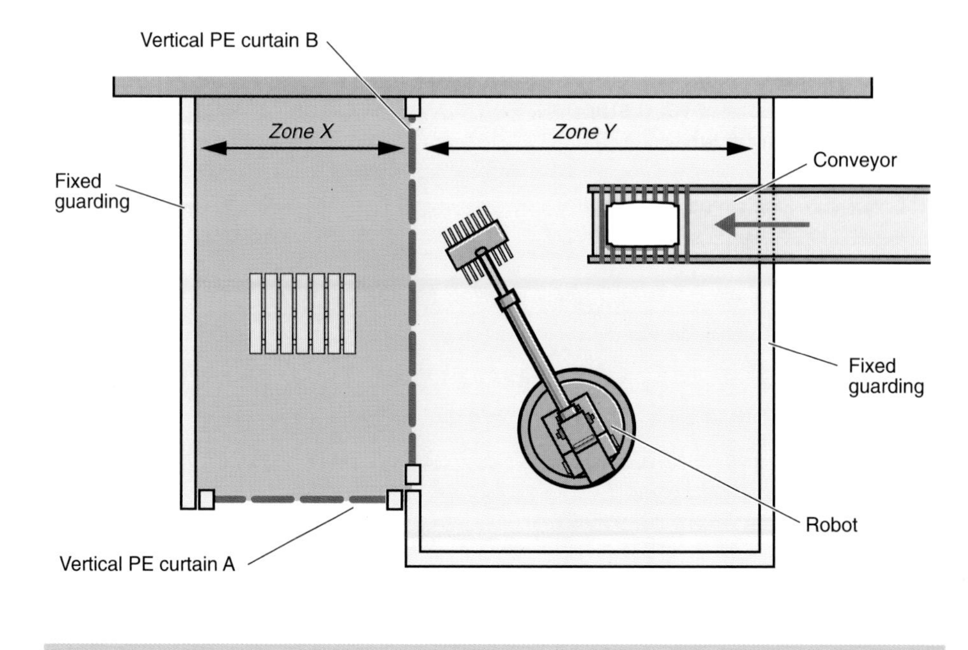


Figure 1 Plan view of a palletising robot safeguarded by light curtains

136 In this mode, light curtain B is active and remains active until the normal control situation is restored manually, ie after Zone X is cleared. The positioning of light curtain B in relation to Zone X takes into account the stopping speed and distance of the robot.

137 In Figure 2, the system protects access to a robot engaged in a simple handling or demonstration task. The vertical light curtain acts as an initial trip device to place the robot on servo-hold. If anyone then tries to get close to the robot, a presence-detecting (horizontal) light curtain will remove power from it. The horizontal light curtain is positioned below the robot arm to permit movement during normal operation.

138 Light curtains are useful where close observation or an unobstructed view of the robot or associated machine is required. They allow zoning for human access and conversely can restrict robot movements at particular times of a robot cycle.

139 There are some factors which make light curtains unsuitable for safety applications such as harsh environmental factors, eg excessive vibration, dust, heat or radiation. It should also be remembered that a light curtain cannot protect against an ejection hazard, eg of tools or workpieces being handled by the robot. Machinery with long stopping times, or situations where a tangible barrier is required (eg adjacent to a walkway) also make a light curtain unsuitable.

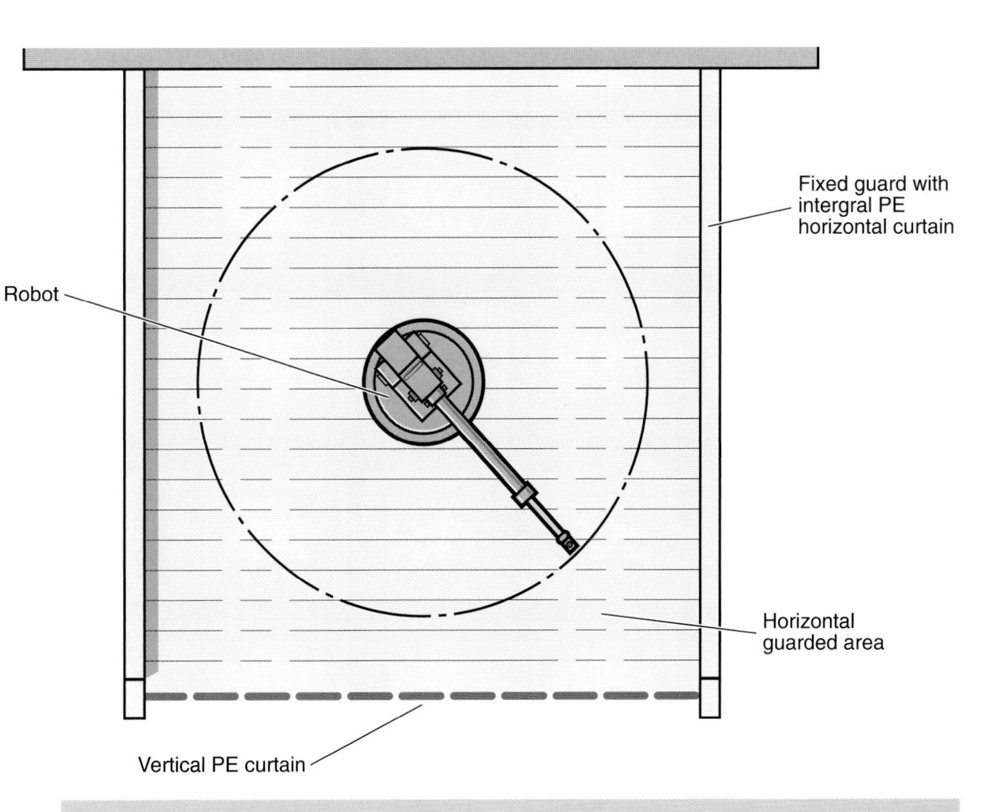


Figure 2 Plan view of demonstration robot safeguarded by light curtains

# Laser scanning devices

140 These are single devices which utilise a scanning laser beam to view and map an area. The sensing zone is two dimensional and programmable so as to recognise any stationary equipment within the scanning zone (area being protected). It detects changes within the programmed map and, in a similar way to a light curtain, it provides a safety output signal to an associated machine interface.

141 In a robot application, a scanner is normally positioned below the working range of the robot(s) but sufficiently high above floor level to detect persons moving into the safety detection zone. It is an alternative method to the use of pressure-sensitive mats. Figure 3 shows how a scanner is typically installed. The PE curtain detects uncovenanted intrusion of the robot into the guarded area and generates an emergency stop of the robot. The PE curtain is muted when the operator initiates the manufacturing sequence after withdrawing from the guarded area.

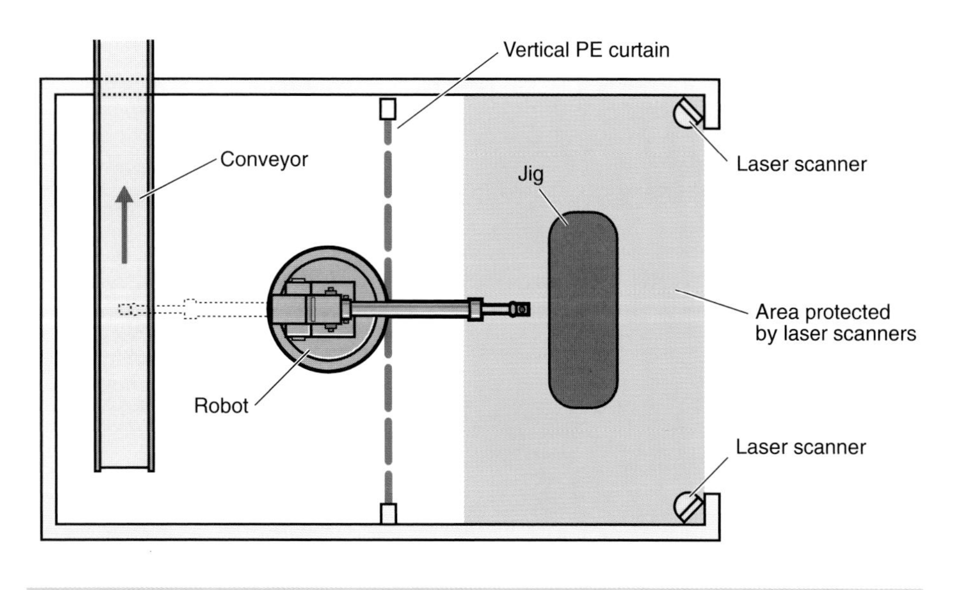


Figure 3: Plan view of robot safeguarded by laser scanner and light curtain

# Capacitance safety devices

142 Capacitance safety devices are considered to be low-integrity devices. They make use of a capacitive loop connected electrically to a control unit which utilises a tuned high-frequency oscillator. The detecting loop is located at a suitable distance from the dangerous parts of machinery. Disturbance of the electric field around the loop by any person's intrusion into the field, causes the control unit to respond by providing a safety stop signal. They can be used as a trip device or presence sensor. Care should be taken to make certain that the correct level of sensitivity is achieved so that the dangerous movement is stopped before the hand or other part of the body tripping the device reaches the trapping zone. Regular checking is recommended to ensure that an adequate level of sensitivity is maintained.

# Other safety measures

## Pressure-sensitive mats and floors

143 These are devices with a sensitive upper surface so that the pressure applied by a person standing on them will cause dangerous motion to stop. The dimensions and the positioning of the mat should take account of a person's speed of approach, length of stride, and the overall response time of the safety system (see BS EN 999). Care should be taken to ensure that access cannot be gained without actuating the mat. Pressure-sensitive mats tend to be insensitive at the edges. Where a number of mats are used together, they should be installed so that adjoining inactive areas do not provide the operator with a pathway to any hazard.

144 Pressure-sensitive mats may be appropriate in circumstances where the use of a fixed guard or an interlocking guard is impracticable. They are particularly suitable as presence-sensing devices, eg to protect a person who may be inside machinery, or in conjunction with other forms of safeguard. However, they have similar disadvantages to light curtains in that they do not protect against ejection hazards. In addition, pressure-sensitive mats are exposed to potential damage which can result in their failure to operate and are not generally considered suitable in situations where high integrity of the safeguard is required (see BS EN 1760 Part 1).

## Two-hand control devices

145 Two-hand control devices offer a means of protecting the operators hands where guarding is impracticable. They may also be used as a hold-to-run control and are therefore generally used in combination with other safeguards. These devices are not suitable where persons other than the operator are at risk. Safety criteria for the design of two-hand control devices are set out in BS EN 574.

## Trip devices

146 Trip devices such as pressure-sensitive edges, pressure-sensitive bars and tripwires can be used to stop the robot if it comes into contact with people (or associated machinery). Care should be taken to ensure that all parts of the robot creating the hazard are protected by a trip device which arrests all motion before a hazardous situation is created. When tripped, the control system should require manual resetting before hazardous motion is possible. Because of the difficulty of applying such devices to the robot arm, they are not normally used for the protection of people.

## Positive stops

147 Positive stops can be used to limit the movement of the robot to that part of its operating envelope required for the current task. This has the advantage that hazards, eg trapping, impact etc may be eliminated and robot movements

kept within well defined limits. These positive stops should be designed or shrouded so that they do not create additional trapping points. Normally they are designed as an ultimate limit on movement if the functional software-based stop fails.

## Brakes

148 Brakes should be provided where there is the danger of gravity fall of a robot arm due to removal of power. It should be capable of supporting the weight of the robot arm and the weight of the largest tool or workpiece likely to be used. Brakes may also be used as a back-up system to other forms of safeguarding, in which case forces created by dynamic energy should be taken into account.

## Emergency stop actuators

149 Emergency stop actuators should be red with yellow background if practicable (see BS EN 418).

150 Emergency stop actuators should be provided at the control panel, on the teach control pendant, at every workstation and at any other prominent position considered necessary. They should comply with BS EN 418; in particular, they should require manual resetting. Buttons should be of the mushroom-headed type. If pull-wires are used they should also comply with BS EN 418.

151 Each emergency stop should be hardwired to the robot power supply and should stop all motion and, where appropriate, release all stored energy. In some robot systems a controlled shutdown may be necessary.

## Enabling devices

152 An enabling device is defined in BS EN 292 Part 1 as: 'additional manually operated device used in conjunction with a start control and which, when continuously actuated, allows a machine to function.' An enabling device is additional to and used in conjunction with other manual controls, particularly hold-to-run controls. It is manually operated, and when continuously actuated, allows machine controls such as hold-to-run controls to function. Until it is actuated, the machine will not be capable of operation, even when start or hold-to-run controls are pressed. The machine will stop immediately the enabling device is released, even if other controls such as hold-to-run are still being operated. Its main use in robot applications is on teach pendants, where it can take the form of a pad gripped by the hand holding the pendant, or pressed by the palm of the hand which is operating the push-buttons. It may also take the form of a three-position switch, gripped in one hand, so designed that releasing the actuator or squeezing it tightly causes the machine to stop.

# Change of state controls

153 Most robots will be used in two quite distinct modes of operation - run and teach/setting. It should not be possible to change the control from one mode to another with personnel inside the safeguarding boundary. The exact way of achieving this controlled changeover will vary depending on the complexity of the installation but the basis of all systems is some form of lock to fix the control mode. This may be a simply key switch, part of a trapped key system or a highly formalised personal padlock system with each person locking off the motive power before entering the enclosure or safeguarded area.

# Slow speed

154 Many robot installations require close approach for teaching/setting etc. This can only be achieved safely by limiting the speed of the robot arms to such an extent that:

- the energy of any impact is insufficient to cause significant injury; and
- the speed is slow enough to minimise the risk of trapping.

155 It should be appreciated that people can be trapped even by slow moving machinery and also that the slow speed cannot be guaranteed under fault conditions. Slow speed should be used in combination with other safeguards to minimise the risk. While it is difficult to give precise information on acceptable speeds, it is unlikely that speeds in excess of 0.25 m/sec will be acceptable. This speed limitation should apply to any part of the robot arm - not just the end effector. It is essential that changing from run to slow speed is achieved under known and controlled conditions (see paragraphs 68-71 on programming and the 'Interfacing with the robot controller' section for ways of achieving this).

## Speed and current sensors

156 Speed and current sensors are devices which sense any abnormalities in robot speed or power and can be used as a hardware safety back-up to the normal position data used in the robot controller.

# Defined areas

157 In some robot applications the motion of the arms creates no hazard because the normal speed is slow and hazardous tools and fixed objects which could create a trapping risk are absent.

158 It is essential that the area swept by the robot arm should be kept clear of objects which, however transitorily, may create trapping hazards. The swept area should be indicated and/or personnel warned when they are entering or inside the swept area. This can be achieved by:

- defining the area with a hatched or painted section of the floor; or
- handrails; or
- a visual or audible alarm activated by:

  - a light curtain; or
  - a pressure-sensitive mat or pressure-sensitive floor; or
  - a laser scanning device.

The exact choice would depend on the location of the robot, frequency of access and type of use.

## Safe systems of work

159 Safe systems of work may be required to further reduce residual risks following risk reduction by design and safeguarding. They should provide a method of working which will ensure the safety of those working with robots and associated machines and equipment. Formulation of a safe system of work will involve identification of hazards, modes of operation, frequency of approach and type of activity. These systems of work should be recorded.

160 It may be necessary to introduce a formal permit-to-work system which should set out:

- a clear handover procedure;
- what work is to be done;
- who is to carry it out and the equipment necessary for the task;
- what safety precautions are to be taken;
- what state the machine is in;
- the expiry date and time of the permit; and
- a clear hand-back procedure.

# Interfacing with the robot controller

161 The electrically based safeguards, described in the 'Safeguarding methods' section, can operate in association with the robot controller. A general description of some methods is given below to illustrate the safety principles involved. The electrical safeguarding methods will form part of the overall safety package.

162 The design of effective safeguards for robot installations is based on the presumption of robot failure, or failures in control systems which could lead to dangerous uncovenanted or unexpected movement of the robot.

163 The principles described deal with the safety control of a robot application through its controller. The controller needs to interface with safety systems and the control systems of associated machines and equipment.

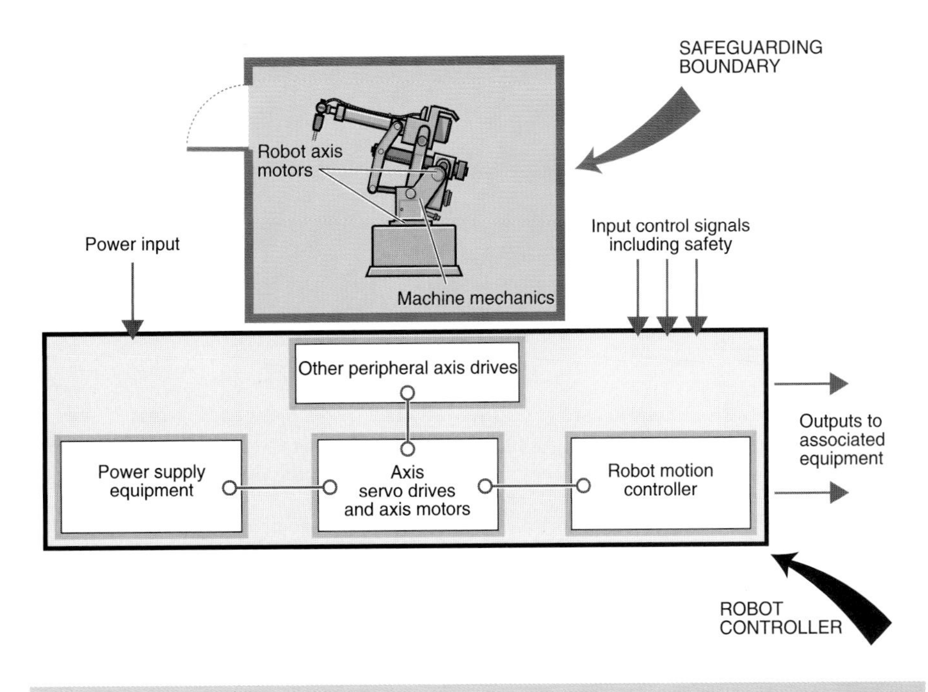


Figure 4 Generalised diagram of robot controller indicating principal features

# The elements of robot systems

164 The elements of an industrial robot are:

- the basic mechanical structure;
- the actuating mechanisms, eg axis motors, hydraulic actuators; and
- the robot controller comprising:
  - power supply equipment which provides the main power supply for the axis servo drive motors and the robot motion controller;
  - axis servo drives which provides the controlled power output, from the robot controller to each individual axis motor; and
  - the robot motion controller which manages the dynamic motion of the robot. It may also manage the motion of turntables or other manipulators which are used in synchronisation with the robot.

These elements can be represented as shown in Figure 4.

165 A typical robot axis servo drive configuration, as shown in Figure 5, is a closed loop position control system, where the automatic control of speed, position and acceleration by closed loop feedback is also illustrated. All practical systems have this facility. Automatic control is achieved by obtaining a feedback signal from a position transducer attached to the drive motor. Some very early robots, which utilised Stepper Motors as the axis drives, did not have a closed loop control system. Robot control systems are emerging that do not include a feedback transducer but derive the control information electronically from the drive motor. This is known as sensorless feedback.

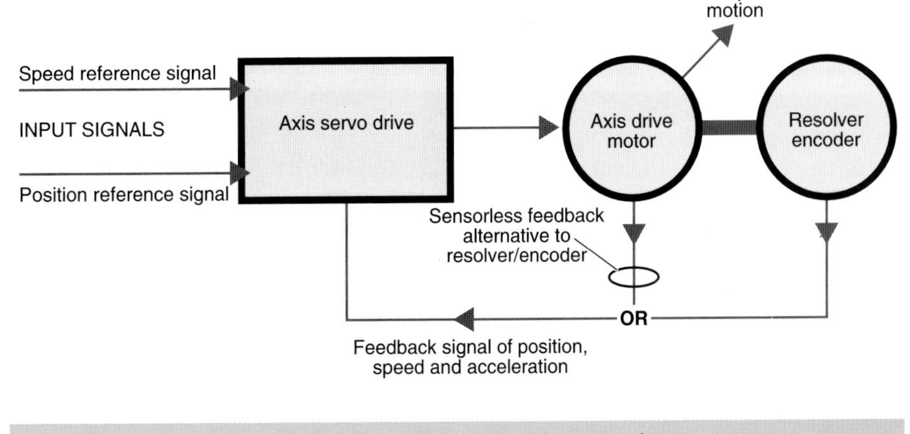


Figure 5 Electric robot: closed loop control

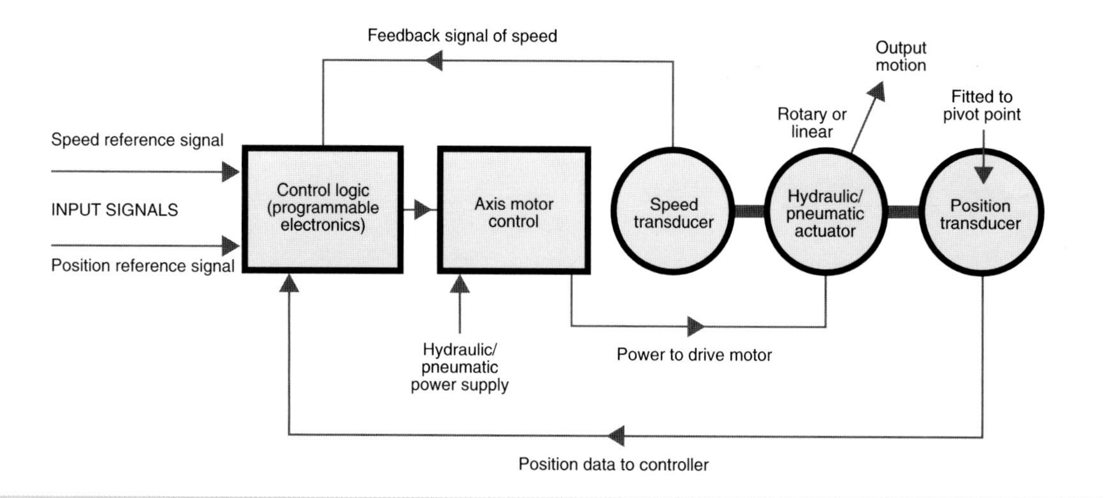


Figure 6 Hydraulic/pneumatic robot

166 Axis motive power can be electric, hydraulic or pneumatic. The servo system for electrical robots is shown in Figure 5 and that for hydraulic/pneumatic in Figure 6.

# Safeguarding methods and their implementation

167 In order to meet the safeguarding requirements, it is necessary to be able to stop or slow down the robot and any appropriate associated machines and equipment or, where practicable, reduce the output power available at the axis drive motors.

## Safety integrity

168 Safety integrity criteria are given in the standards which support the Machinery Directive 98/37/EC. The relevant standards are itemised in the References section.

169 An important element in the implementation of an electrically based safeguarding scheme is the need to determine the level of safety integrity that has been achieved and then to assess whether this is acceptable for the hazards and risks in question. The term safety integrity refers to the ability of a safety-related system to perform its safety functions correctly. In determining the safety integrity, all causes of failure which could lead to an unsafe state should be included, eg hardware failures, software failures and failures caused by electrical interference.

170 The overall assessment of the safeguarding scheme should be carried out systematically and should include the following steps:

**Step 1:**

Identify the hazards and determine the events leading to specific hazards, for example teaching mode, uncovenanted motion of the robot or associated machines and equipment

**Step 2:**

Determine the required levels of safety integrity taking account of severity of injury, frequency of access etc (see BS EN 1050 and BS EN 954-1)

**Step 3:**

Design the safety-related systems using the safety integrity criteria for the robot application, ie arc or spot welding, water jet cutting etc

**Step 4:**

Ensure specified safety integrity level has been achieved

Note: In this context, safeguarding means the safeguarding equipment external to the robot itself comprising the interface components, the hardwired equipment connected to the interface, and the safeguarding devices.

## Robot recovery

171 This is a feature of the robot which relates to the way the robot will move after it has been restarted following a shutdown by, for example, the safeguarding system. Information on these features is normally provided in the manufacturer's written specification for the robot.

172 When designing the installation safeguards, it is particularly important to take account of automatic movement of axis components when they are returning to their datum position. Modern robots have restart software methods for returning the moving parts, via the route they have taken, to a previous known position.

173 Danger from robot recovery is reduced if the robot is brought to rest in a controlled manner. This philosophy is followed in Method 2 (see paragraphs 183-185). In the sections that follow, each safeguarding method is considered from both an implementation and a safety integrity viewpoint.

## Programmable or complex electronic hardware versus hardwired systems

174 An important feature of any electrically based safeguarding system is whether the safety functions are hardwired without complex electronics, or **depend** on programmable or complex electronics. This has important implications for the ease with which the safety integrity can be established.

175 Typical hardwired systems are those which include electro-mechanical relay logic and systems based on discrete electronic components or non-programmable integrated circuits.

176 Interlocking methods utilising hardwired systems are usually less complex and the determination of the level of safety achieved is easier than for systems incorporating programmable electronics. A useful strategy is therefore to exploit the power and flexibility of the programmable or complex electronics yet ensure that an adequate level of safety has been achieved by hardwiring the safety functions.

177 This approach was used in the design of older installations but for technical and economic reasons modern installations include software-based safety systems designed to have an equivalent level of safety integrity as that of a hardwire-based system.

178 When users of robots are designing an installation, it is recommended that the relative merits of both systems described in the previous two paragraphs are considered.

179 A number of electrically based safeguarding methods are described in the following paragraphs. These examples are not exhaustive but may provide a basis for other designs.

180 The aim is to illustrate a wide range of safeguarding strategies which may depend on programmable electronic systems (PES) or on non-PES arrangements. When the safeguarding strategy is based on non-PES arrangements, the safety integrity of the system may be enhanced by utilising electronic monitoring systems which could be separate from or part of the robot controller.

## Method 1 Robot controller - direct hardwired

181 This method is particularly appropriate for older robot designs. In Figure 7, the operation of the actuation device (eg guard position detection switch) removes, by control means (path 1a), the power to all the axis drive motors via a disconnecting device (eg a motor contactor) in the axis servo drives (path 2). The robot motion controller is also informed that the power to each axis has been removed (path 1b); this method is independent of the robot motion controller. It should be noted that even though the programmable electronics are informed that the power is to be removed from the axis motors, there may be problems with this method in relation to robot recovery.

182 In Figure 7 the request access device is a guard position detection switch mounted directly at the gate in the safeguarding fence. However, it is not a requirement of this method that the request access device be a guard-mounted device. Installations may be designed with a request access device located at a position convenient to the operator. This enables the release of a mechanical key with which the gate may be subsequently opened via a key-exchange system. If this method is used, it should not be possible to remove the key which allows the robot to operate while the gate remains open. This option eliminates the need for vulnerable electrical cables, which are required by fence-mounted safety devices such as guard position detection switches and guard-mounted electro-mechanical gate-locking devices.

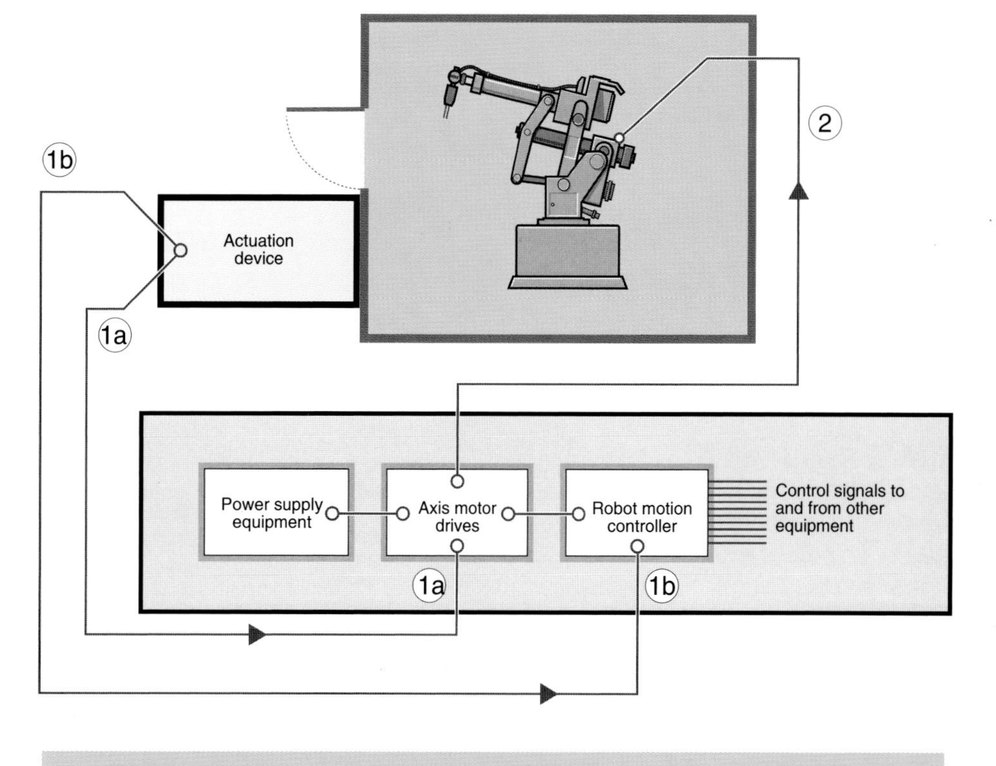


Figure 7 Method 1: Direct hardwired

# Method 2 Robot controller with hardwired back-up

## *Method 2.1 Interactive back-up, captive key switch*

183 The essential feature of this method is an actuation device, eg a request access push-button plus a switch which incorporates a locking solenoid. In Figure 8 the actuation device is located convenient to the operator and the captive key switch mounted at the access gate to the enclosure.

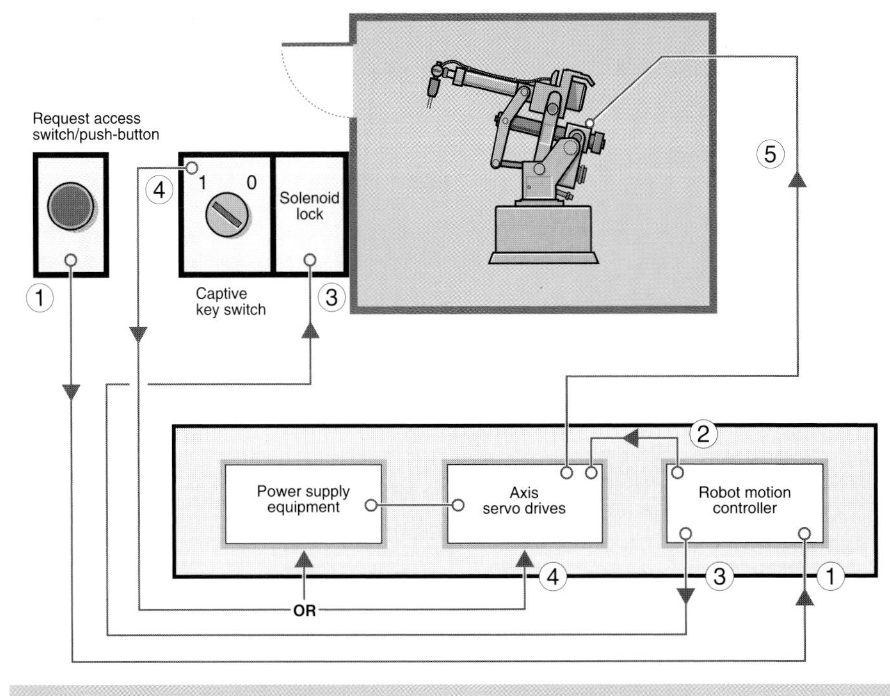


Figure 8 Method 2.1: Interactive back-up captive key

184 Access to the safeguarded enclosure is requested by operating the push-button. This causes a controlled shutdown of the robot through the robot motion controller (paths 1 and 2). The axis drive motors are consequently held stationary (in the servo-hold state) through the robot motion controller. A signal is then sent from the robot motion controller to the captive key switch (path 3) to release the solenoid locking device so that the key can be turned to position 1.

185 Rotation of the key to position 1 unlocks the access gate and sends a signal (path 4) to a means of disconnection (eg a contactor or isolator) in the axis servo-drives or power supply equipment. This will remove the power to all the axis drive motors (path 5). The key is held captive in the gate switch and can only be removed when the gate is locked shut.

## *Method 2.2 Interactive back-up, trapped key switch*

186 In Figure 9 the actuation device is a request access push-button plus a switch that incorporates a locking solenoid, both of which can be mounted in a location convenient to the operator. The method of operation is as follows.

187 Access to the safeguarded enclosure is requested by operating the push-button. This causes a controlled shutdown of the robot through the robot motion controller (paths 1 and 2). The axis drive motors are consequently held stationary (in the servo-hold state) through the robot motion controller. A signal is then sent from the robot motion controller to the captive key switch to release the solenoid locking device so that the key can be turned to position 1 (path 3). Rotation of the key to position 1 sends a signal (path 4), to a means of disconnection (eg a contactor or isolator) in the axis servo-drives or power supply equipment. This will remove the power to all the axis drive motors (path 5). The key can then be removed from the trapped key actuation device and used to unlock the gate to the enclosure. This key is then trapped in the gate switch and can only be removed when the gate is locked shut.

Methods 2.1 and 2.2 are **not** recommended where whole-body access is required to the enclosure.

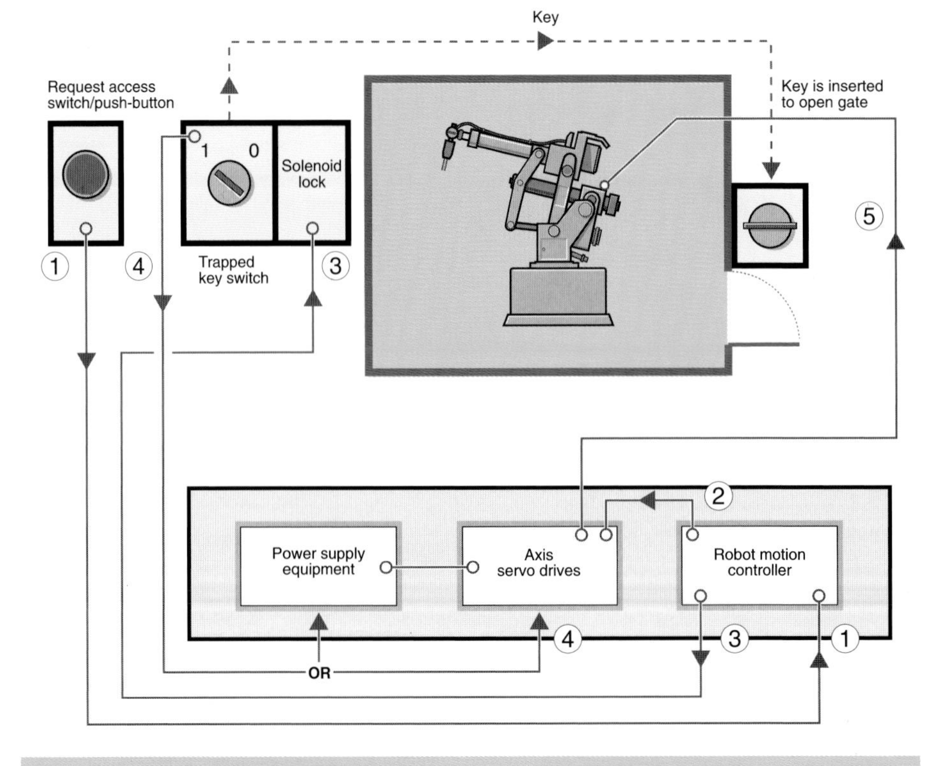


Figure 9 Method 2.2: Interactive trapped key - hardwire reliant

### *Method 2.3 Interactive back-up, trapped key switch with key exchange*

188 The access procedure shown in Figure 10 is similar to that in Figure 9 except that in this case the key is removed from the trapped key actuation device and inserted into a gate-locking key-exchange device mounted on the access gate to the enclosure. Rotating the key in this second device traps it, unlocks the gate, allows the second trapped key to be removed from the gate-locking device and simultaneously causes either:

- removal of the power to all the axis drive motors by a disconnecting means (eg a contactor) in the axis servo-drives (path 7); or
- removal of the power to all the axis servo-drives by direct mechanical means (eg an isolator) in the power supply equipment (path 6).

Power is thereby removed from all axis drive motors (path 5). The person entering the enclosure removes the second key from the gate-locking device and retains it. This prevents the robot being started inadvertently when someone is inside the enclosure. The method described in 2.3 may be used where whole-body access is required to the enclosure.

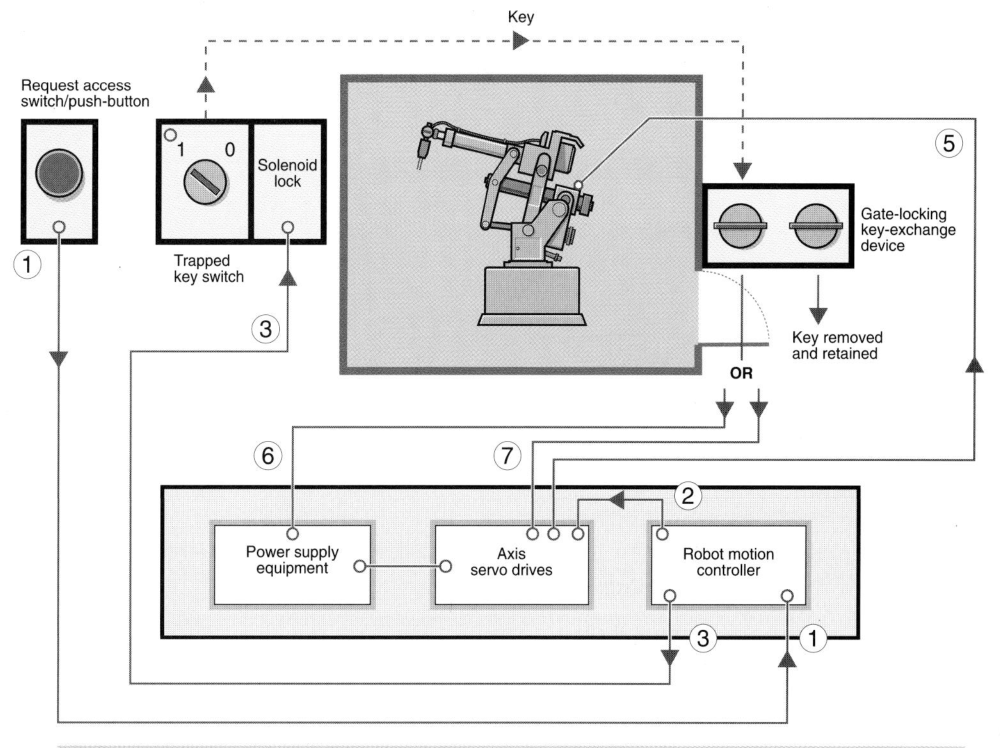


Figure 10 Method 2.3: Interactive trapped key - hardwire reliant key-exchange system

## Method 3 Interactive trapped key - software reliant

189 In Figure11, access to the safeguarded enclosure is requested by operating the push-button. This causes a controlled shutdown of the robot through the robot motion controller (paths 1 and 2). The axis drive motors are consequently held stationary (in the servo-hold state) through the robot motion controller. A signal is then sent from the robot motion controller to the actuation device to release the solenoid-locking device so that the key can be turned to position 1 (path 3). Rotation of the key to position 1 enables the key to be removed and inserted into the gate-locking key-exchange device. Inserting the key and unlocking the gate traps the key in position and releases the second key, which is removed and retained by the person entering the enclosure. This prevents the robot being started inadvertently when someone is inside the enclosure.

190 In such applications there is no hardwired means of disconnection in the axis servo-drives and it is not normal practice to use the mechanically operated means (an isolator) in the power supply equipment. Safety integrity relies on the inherent monitoring features of the robot motion controller and the integrity of the software. Paths 1, 2, 3 and 5 in Figure 11 illustrate this method. If this method is used, the designer and/or supplier would be expected to demonstrate a high integrity in the performance of the safety functions.

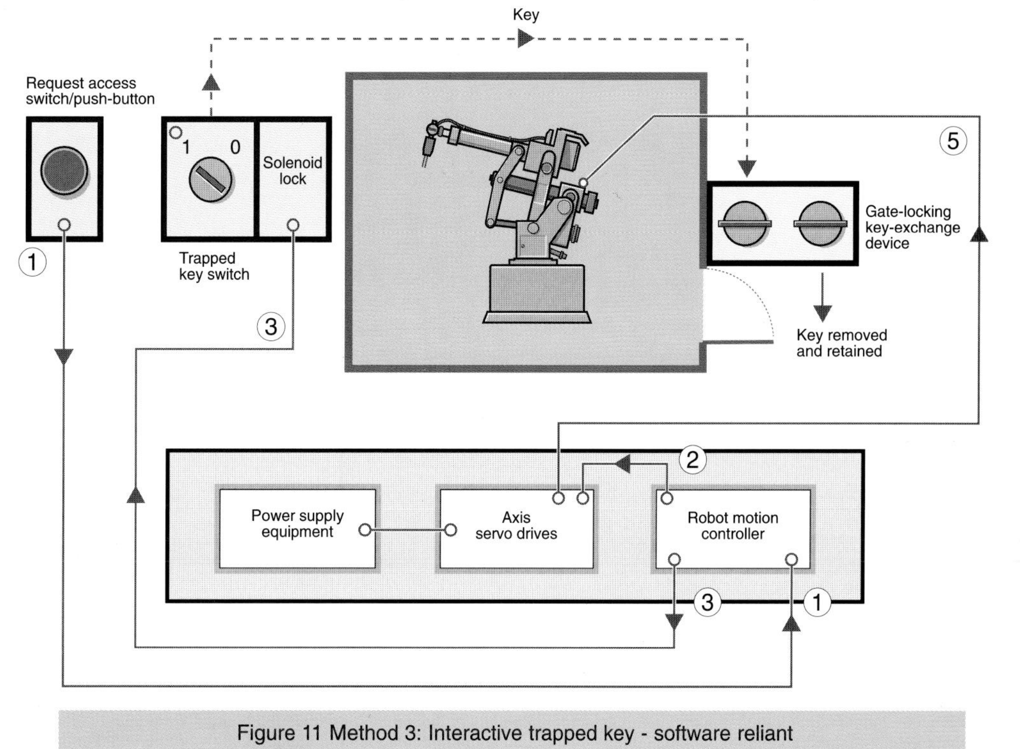


Figure 11 Method 3: Interactive trapped key - software reliant

# Method 4 Independent back-up

191 This is shown in Figure 12 and is relevant to older designs, particularly hydraulic robot drives. The essential feature of this system is a timer which is set to the time it takes:

- for the robot motion controller to effect a controlled shutdown of the robot at a known safe point in the cycle; and
- to allow rotating tools, eg grinding wheels, to run down.

When the set time has elapsed, the timer:

- sends a signal to a means of disconnection (eg a contactor or isolator) in the axis servo-drives or power supply equipment (path 3); and
- enables the key switch to be turned to position 1 (the gate unlock) position.

192 If the key switch is gate-mounted, turning the key will unlock the gate directly and release the key for withdrawal and retention by the person entering the enclosure. If the key switch is located in a position convenient to the operator, turning the key will release it for withdrawal. It can then be removed from the key switch, used to unlock the gate mechanically and retained by the person entering the enclosure.

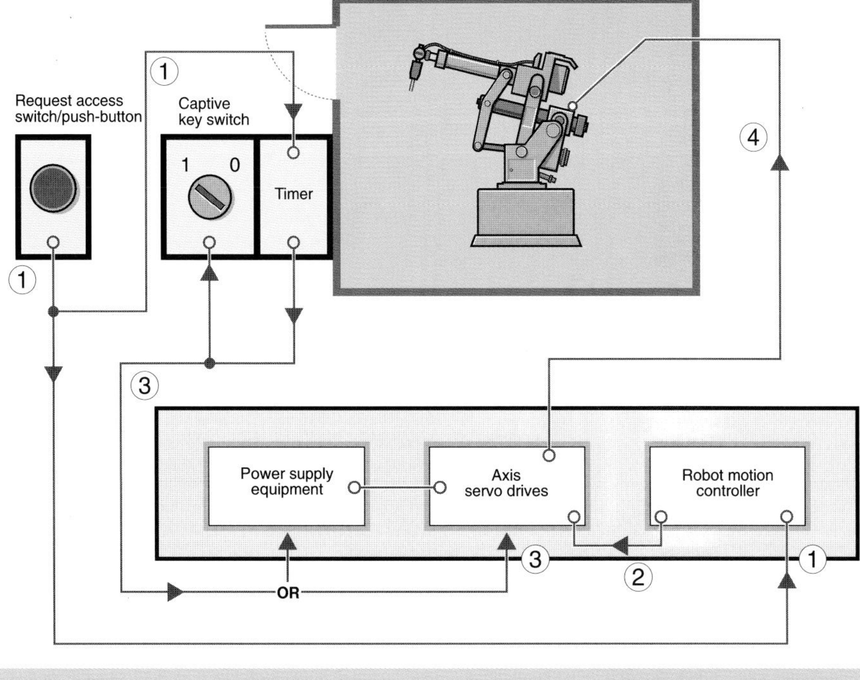


Figure 12 Method 4: Independant back-up timer

# Method 5 Robot controller with external safety system monitoring circuit

193 In this method an external safety system monitors critical safety functions and puts the robot system into a safe state if a failure of those functions being monitored is detected. This method could form part of the overall safety package used, for example, during teaching and maintenance.

194 There are a number of design variations within the scope of this method. In Figure 13 the operation of the actuation device puts the axis drive motors into servo-hold via the programmable electronics (paths 1, 2 and 5). On receipt of the signal from the actuation device (path 1), the monitor checks that the robot is maintained in this mode. Deviation would be detected by the monitor which would remove, by control means in the power switching control (path 3), the power for all axis motor control modules and therefore the power to all axis drive motors (paths path 4 and 5). The system can be arranged so that it monitors critical functions, eg robot speed during teaching (from the teach pendant within the robot enclosure) to ensure that the robot does not exceed the reduced speed.

195 The safety integrity criteria will depend on the degree of reliance being placed on the monitor and factors such as whether the monitor is hardwired or based on programmable electronics. Although the actuation device is shown in Figure 13 as mounted at the gate, similar options for the type and location of the request access switch (and gate-locking system) apply as in previous methods.

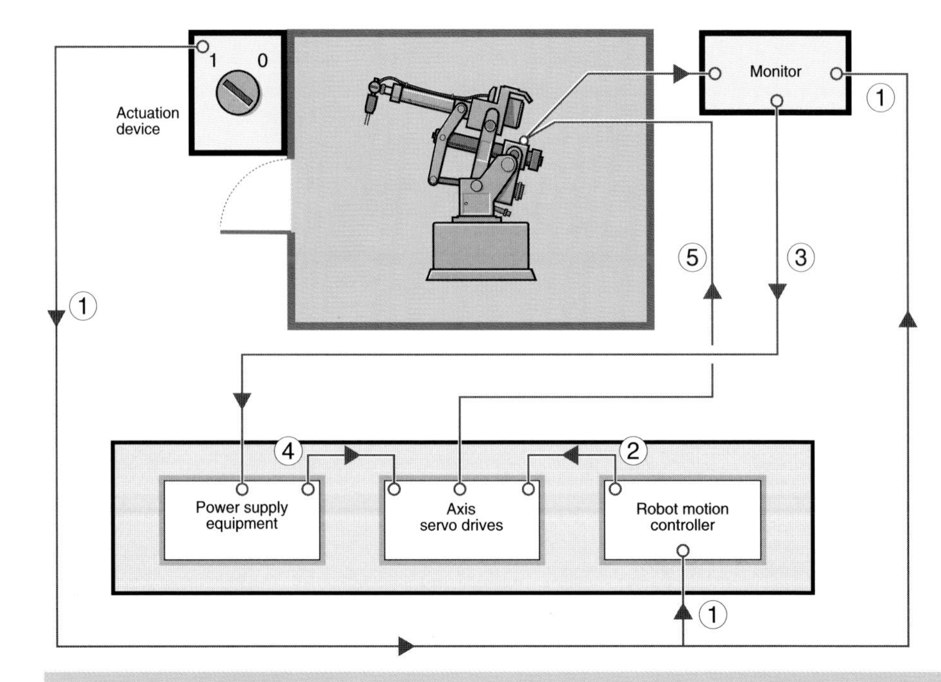


Figure 13 Method 5: Robot controller with external safety system monitoring circuit

# Method 6 Robot controller with a programmable electronics back-up

196 In Figure 14 the safeguarding functions are carried out by the robot controller and separate programmable electronic equipment which acts as a back-up (path 3). The request access switch causes:

- the programmable electronics of the robot controller (path 1) to effect a controlled shutdown of the robot, putting the axes drive motors in servo-hold (paths 2 and 6); and
- the separate programmable electronics to remove the motive power to each robot axis after a specified delay (paths 4, 5 and 6). The delay should be sufficient to allow the controlled shutdown of the robot by the robot controller. The duration and safety implications of this delay will need to be taken into account at the design stage.

197 The separate programmable electronics could be located within the robot controller, and form an integral part of it, or be located outside it as a separate piece of equipment such as a programmable logic controller performing various supervisory functions as part of a machining cell. Although the actuation device is shown in Figure 14 as mounted at the gate, similar options for the type and location of the request access switch (and gate-locking system) apply as in previous methods.

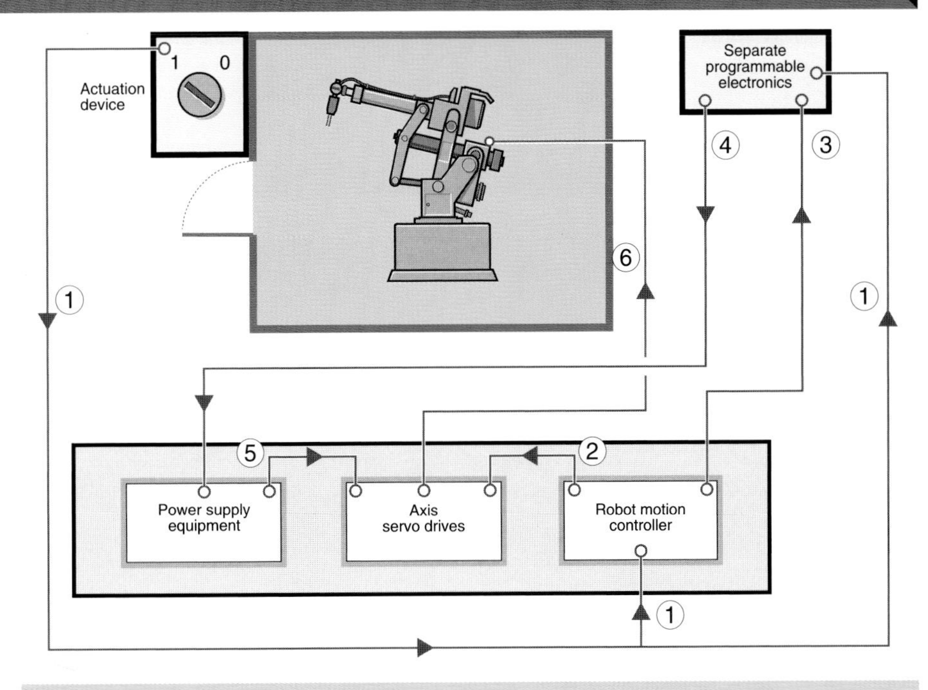


Figure 14 Method 6: Robot controller with programmable electronics back-up

# Teaching - methods of safeguarding

198 Typical relevant electrical features are represented in Figure 15 and comprise:

- axes-reduced speed control;
- hold-to-run teach controls - these are implemented by the robot motion controller, which is capable of effecting robot motion when the teach buttons are depressed and stopping robot motion when the buttons are released. Joysticks are an alternative to push-buttons;
- an independent emergency stop;
- an enabling device (usually part of the hand-held teach control but may be a separate hand-held device); and
- a display screen which shows a selected condition and questions the user regarding their desire to continue.

199 Operation of an independent emergency stop causes removal, by control means in the power supply equipment, of the power to all axis drive motors. Unless the enabling device is continuously actuated, the hold-to-run controls are inoperative. Release of the enabling device causes removal, via the robot motion controller, of the power to all the axis drive motors. Figure 15 shows the basic and necessary features that are required.

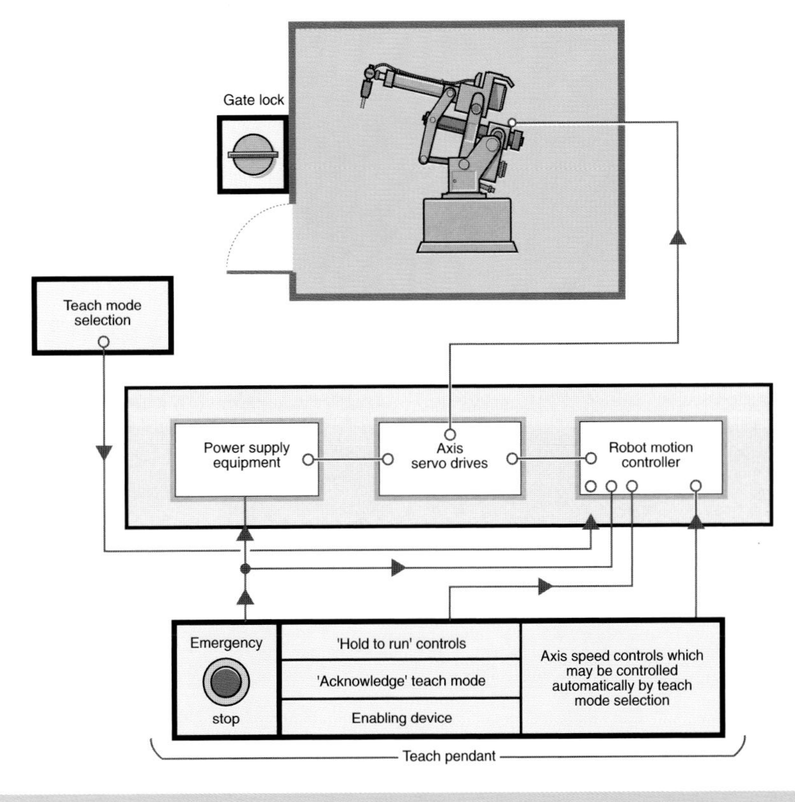


Figure 15 Teaching - typical features

# Appendix 1 Health and safety law

1 This appendix outlines the main legal requirements that relate to the safety of industrial robot systems - there are full details of all the relevant legal publications in the References section. Manufacturers, suppliers, employers, the self-employed and consultants all have legal obligations which are imposed by the Health and Safety at Work etc Act 1974 (HSWA) and its relevant statutory provisions. These requirements are enforced by either the Health and Safety Executive or by local authorities, depending on the premises, as determined by the Health and Safety (Enforcing Authority) Regulations 1998.[8]

## Health and Safety at Work etc Act 1974 (HSWA)

2 Section 2 of the Act places general duties on all employers to ensure the health, safety and welfare of their employees so far as is reasonably practicable. These duties include:

- the provision and maintenance of plant and systems of work that are, so far as is reasonably practicable, safe and without risk to health;
- the provision of places of work and access and egress to them that are safe and without risks to health; and
- the provision of adequate information, instruction and training as is necessary to ensure employees' health and safety.

3 Section 3 places general duties on employers and the self-employed to take care of their own and other people's health and safety where this may be affected by the employer's work activity. This section therefore addresses the work activities of consultants who may be retained to provide technical advice.

4 Section 6 places general duties on designers, manufacturers, installers, importers or suppliers of articles for use at work to:

- ensure, so far as is reasonably practicable, that articles are designed, manufactured and installed so that they are safe and without risk to health;
- carry out such testing and examination which is necessary to ensure compliance with these duties;
- provide users with the necessary information to ensure the safe use of those articles; and
- take the necessary steps to ensure, so far as is reasonably practicable, that such information is revised and provided to users when anything becomes known which could give rise to a serious risk to health and safety from use of the article.

## Supply of Machinery (Safety) (Amendment) Regulations 1994 (SMSR)

5 The Supply of Machinery (Safety) Regulations 1992 (as amended) (SMSR) place duties on all persons who supply relevant machinery to ensure that it is safe. Suppliers must ensure that the relevant essential health and safety requirements (EHSRs) for the machinery are satisfied, taking into account the principles of safety integration (Requirement 1.1.2).

6 Requirement 1.2 of the EHSRs (listed in Schedule 3 to the Regulations) is of particular relevance to robot control systems. It addresses the following issues:

- safety and reliability of control systems, including errors in logic;
- control devices;
- starting;
- stopping devices including normal stopping and emergency stop;
- mode selection (eg teaching mode and end-effector changing etc);
- failure of the power supply;
- failure of the control circuit; and
- user-friendly software.

7 SMSR requires the drawing up of a technical file, the issuing of an EC declaration of conformity and CE marking if the machinery meets relevant EHSRs and is safe. For relevant machinery which cannot function independently and is intended for incorporation in a larger assembly, a declaration of incorporation should be issued by the supplier but the machinery should not be CE marked. In these situations the end supplier, which may be the user, is responsible for ensuring that the final assembly complies with the Regulations.

8 SMSR applies to all new relevant machinery when first supplied after 1 January 1995. Second-hand machinery supplied for the first time within the European Economic Area after this date must comply with the requirements of SMSR, irrespective of the original manufacture date.

9 In some cases, other supply legislation may apply such as the Electrical Equipment (Safety) Regulations 1994 (see page 32) or the Electromagnetic Compatibility Regulations 1992. The existence of CE markings on machinery should indicate that the manufacturer has met all of the supply requirements that are relevant.

# Management of Health and Safety at Work Regulations 1999 (MHSWR)

10 Employers have duties under these Regulations which include:

- the assessment of risks to the health and safety of employees and non-employees and to review the assessment when there is significant change in circumstances;
- arrangements for the effective planning, organisation, control, monitoring and review of those preventive and protective measures identified in the risk assessment as being required;
- the appointment of a competent person to assist them in complying with their legal obligations; and
- to restrict access to employees to any area of the workplace, where necessary on the grounds of health and safety, unless employees have received adequate health and safety instruction (such restricted areas are referred to as danger areas).

11 A danger area is a work environment which needs to be entered by an employee and where the level of risk is unacceptable without special precautions being taken. The hazard involved need not occupy the whole area but can be localised, for example where an employee is likely to come into contact with a hazard while 'teaching' a robot system.

12 The self-employed also have duties under MHSWR, including undertaking risk assessments.

# Provision and Use of Work Equipment Regulations 1998 (PUWER)

13 These Regulations place a number of duties on employers, the self-employed (for equipment they use at work), persons in control of work equipment or the way in which such equipment is used, and persons who are in control of the supervision or management of the use of work equipment. These duties include:

- ensuring that risks created by the equipment are eliminated where possible, or controlled by taking appropriate measures;
- ensuring work equipment is constructed or adapted to be safe and suitable for its use;
- selecting work equipment which will be suitable for the working conditions in which the work equipment will actually be used;
- ensuring work equipment is only used in conditions for which it is suitable;
- ensuring that work equipment is maintained in efficient working order and in good repair;
- inspecting work equipment after installation and before use, where the safe operation depends on the installation conditions, to ensure it has been installed correctly and is safe;
- periodically inspecting work equipment exposed to conditions causing deterioration that is liable to result in a dangerous situation;
- inspecting work equipment following exceptional circumstances which may jeopardise the safety of the work equipment;
- ensuring work equipment is safeguarded to prevent risks from mechanical and other specific hazards;
- providing equipment with appropriate and effective controls (including emergency stop) and control systems;
- providing a suitable means to isolate the work equipment from all its sources of energy (ie when maintenance work is to be carried out);
- ensuring that an inspection is carried out by a competent person and that a record is kept until the next inspection; and
- ensuring that the work equipment is used only by people who have received adequate information, instruction and training.

14 Regulation 11 of PUWER is relevant to robotic systems. It requires the safeguarding of dangerous parts of the robot and dangerous motion, by preventing access to dangerous parts or stopping the movement of dangerous parts before any person enters a danger zone.

15 Regulation 11(3) is of particular relevance to robot systems. This states that all guards and protective devices provided shall:

- be suitable for the purpose for which they are provided;
- be of good construction, sound material and adequate strength;
- be maintained in an efficient state, in effective working order and in good repair, and not be easily bypassed or disabled;
- be situated at sufficient distance from the danger zone;
- not unduly restrict the view of the operating cycle of the machinery, where such a view is necessary;

- be so constructed or adapted that they allow operations necessary to fit or replace parts and for maintenance work, restricting access so that it is allowed only to the area where the work is to be carried out and, if possible, without having to dismantle the guard or protection device;
- not give rise to any increased risk to health or safety.

16 The measures which should be taken are placed in a hierarchy, the highest of which is the provision of fixed guarding. Other measures include the provision of other types of guard, protection devices and appliances, and lastly the provision of information, instruction, training and supervision.

17 Risks from activities such as teaching, setting, cleaning, maintenance and repair of the robot may require a different set of measures from those required when the robot system is in routine use.

# Lifting Operations and Lifting Equipment Regulations 1998 (LOLER)

18 These Regulations apply to robots which carry out lifting operations as part of their programmed work cycle, eg in mechanical handling and palletising operations. Your risk assessment, as required by MHSWR regulation 3, should identify significant risks to people arising from failure or deterioration of the robot when carrying out a lifting operation. In these situations you should take measures necessary to comply with the relevant provisions of LOLER.

These are:

- that the robot and any attachment is of adequate strength and stability for each load;
- that the robot is installed so as to reduce as low as reasonably practicable the risk of the equipment or load striking a person and that it is otherwise safe;
- that the information which indicates the safe working load is kept with the robot;
- that every lifting operation of the robot is properly planned and carried out safely;
- that the robot is thoroughly examined by a competent person to ensure it is safe to operate:

  - after installation and before being put into service for the first time;
  - following relocation;
  - periodically, where deterioration is likely to lead to dangerous situations; and
  - after exceptional circumstances which are liable to jeopardise the safety of the robot.

- that written reports of thorough examinations are made as soon as is practicable;
- that any defect in the robot which could become a danger to persons is notified to the employer forthwith and that it is not used before the defect is rectified; and
- that reports of thorough examinations and other relevant information is kept available for inspection.

19 See paragraphs 11-12 of the Introduction to this book for practical guidance on the application of LOLER to robots.

# Electrical Equipment (Safety) Regulations 1994

20 These Regulations[9] place a duty on the manufacturer (or their authorised representative) or supplier to supply safe electrical equipment. In particular, people must be adequately protected against danger of physical injury or other harm which might be caused by direct or indirect electrical contact.

# Appendix 2 Case studies

1 These seven case studies are based on real installations which have been idealised to demonstrate the principles put forward in this guidance. They are intended to illustrate the way safeguards may be selected for particular applications and types and levels of hazard. There is no suggestion that these safeguarding arrangements are the only ones suitable for these types of application.

2 The case studies follow a common format: a general description of the installation, a description of the operation of the installation, an outline of the hazards and the safeguards, and a more detailed description of the hazards and safeguards for particular operations. Case Study 3 includes a hazard analysis.

## Case study 1: Robots in automotive body spot-welding line

### General description of installation

3 This example is a general representation of a car body component welding line as found in most major car manufacturers. It is used to assemble pressed side body components and spot-weld them prior to final assembly on a similar line.

4 The installation shown in Figure 16 consists of a central conveyor (known as the track or line) with a variety of spot-welding robots, both floor- and gantry-mounted, sited alongside it. In this particular line, an additional robot applies sealant prior to final sub-assembly.

5 Because of the need to sustain continuous production and minimise downtime, great attention has been paid to providing good access for maintenance and routine tool adjustment. Provision has also been made for a manual workstation at the end of the track so that incomplete work can be made good. In practice, this facility is rarely, if ever, used.

6 The track and robots are enclosed by a 2 m-high fence. Each bay is separated from the next by further 2 m fences. Access into each bay is via hinged gates interlocked to both the track drive and the robots' drives by a trapped key exchange system. Local control for each robot for teaching via a pendant control can be achieved using the same trapped key system. Side-body components are assembled manually onto a two-sided rotating jig and are held at key location points by manually-operated clamps. Access to the jig is through a vertical light curtain which prevents rotation when the light beams are broken. Rotation and additional power clamping are initiated by a two-hand control sited outside the loading zone. Access to the manual rectification area and the end of the track is controlled by a light curtain operating in conjunction with proximity devices.

### Installation operation

7 The sequence of operations starts with the operator manually loading various presswork sub-assemblies onto a rotating jig. When these have been assembled and clamped, the operator rotates the jig so that it faces into the enclosure where a robot-mounted spot-welder joins the sections. Meanwhile the operator loads the next assembly onto the reverse side of the jig. When welded, the assembly is automatically loaded onto the track, where it is automatically clamped in position and moves to the next station where a sealant is applied by a robot carrying a sealant gun. The track moves the assembly to the next bay, where additional presswork pieces are automatically positioned and spot-welded by two gantry-mounted robots (the presswork feed conveyor has been omitted from the case study illustration to aid clarity). The final bay contains a second robot sealant applicator. The assembly then leaves the enclosure and passes through a bay which can be used for manual work, if required, before being automatically unloaded from the track (not shown in the case study illustration).

### Hazards

8 The assembly line is a complex arrangement of transfer machinery and robots containing a variety of hazards, many closely related to each other. Only machinery hazards are considered here and these can be itemised as follows:

- trapping between fixed and moving parts of the transfer line as it moves between workstations;

- traps between the workstation clamp and the body panels and/or track as the sub-assembly is clamped in position;
- trapping between fixed and moving parts of the sub-assembly jig as it rotates;
- trapping by the sub-assembly jig clamps;
- traps between the moving robot arms and fixed parts of the installation;
- collision with a moving robot arm;
- traps between the fixed and moving parts of the robot gantry; and
- contact with the robot-mounted welding heads.

9 The installation was the subject of a detailed hazard identification and risk assessment (see paragraphs 109-119 in the main text). This study showed that, for the operators, it was necessary to provide safeguards which allow quick and easy access for loading. For teaching/programming and maintenance it was necessary to provide safeguards which allowed freedom of access within the enclosure while giving a high standard of protection that the close contact with the hazards requires.

## Additional safety arrangements for the operator

### *Hazard: Trapping from the assembly jig*

10 The following safeguards are provided:

- a vertical light curtain in front of the jig inhibits movement of the clamp and rotary motion while it is broken;
- a horizontal light curtain just above floor-level (between the vertical light curtain and the jig) prevents a person standing between the vertical light curtain and the jig. This allows uninterrupted access to load the jig for maximum productivity; and
- a two-hand control device on a pedestal, out of reach of the jig, initiates rotation and power clamping.

### *Hazard: Contact with the track and robots near exit of enclosure*

11 The fence enclosing the assembly line extends to the point where sub-assemblies are transferred to another process. At the end of the assembly line, the following safeguards are provided:

- the exit area is protected by a vertical photoelectric curtain which covers the exit area and detects the presence of an operator. The track will not move if the curtain is interrupted. A proximity device consisting of a pair of inductive sensors on both sides of the light curtain operates in combination with the curtain to allow the free passage of the vehicle body through the opening. As the body section approaches the opening, the first pair of sensors detects the body and mutes the light curtain for its passage. As the end of the body moves through the opening, it is detected by the second pair of sensors which re-enables the light curtain; and
- access to the track apart from the body assembly and clamps is prevented by fixed enclosing guards.

## Safety arrangements for programming/ teaching/ maintenance personnel

### *Hazard: Trapping and collision hazards on the line*

12 The following safeguards are provided:

- a 2 m-high fence encloses the entire area around the track. In addition, each workstation is divided into separate bays by additional 2 m-high fences at right angles to the track;
- access into each bay is via a hinged gate. The gate is mechanically locked shut by a trapped key interlock;
- the power supply to the track and the robots is interlocked to the various access gates as follows:

- key removed from lock which directly isolates the power to the track drive and signals to the robot controllers so that power is removed from each robot and the mechanical brakes are applied;

- the key is then placed in an intermediate exchange box which allows a gate access key to be withdrawn. Each access key is unique and can only open its own particular access gate. No movement of the track is possible until the keys are reinstated in both the intermediate box and the final isolating box;
- the intermediate exchange box also contains trapped keys which can be used at each robot control console to activate local control via the teach pendant system. It is not possible to control the robot other than via the local control until the keys have been replaced in the intermediate box (see paragraph 188 in the main text);
- it is therefore possible to have access to a bay with power to control the robot via the pendant control with the track isolated;

- hardwired emergency stops are situated at a variety of working positions within the enclosure and on the teach pendant;
- the teach pendant is provided with an enabling device and hold-to-run movement controls;
- slow speed only is available while programming; and
- safe systems of work are followed for all foreseeable tasks within the enclosure.

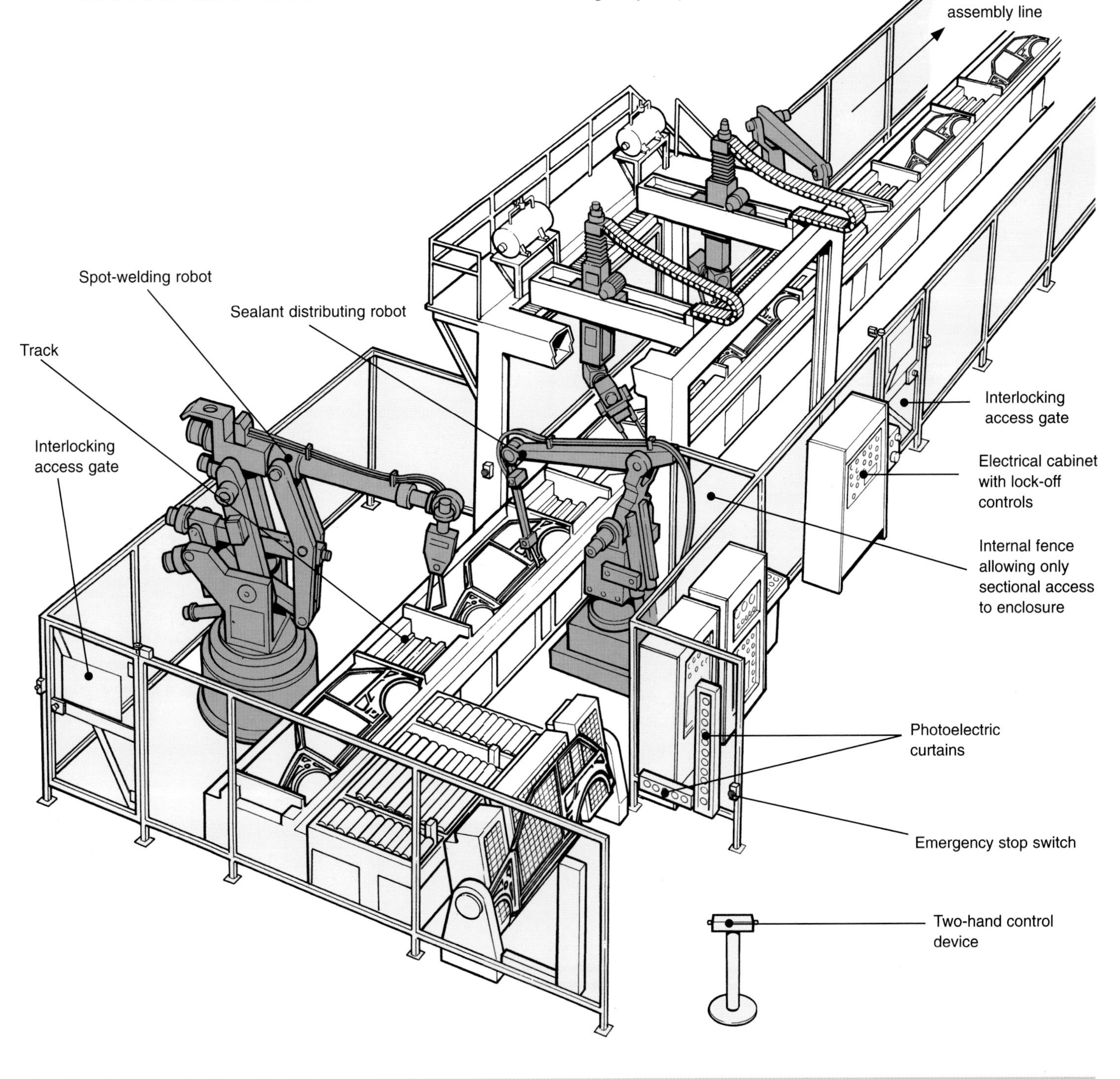


Figure 16 Case study 1: Automotive spot-welding line

# Case study 2: Robot servicing a machining installation

## General description of installation

13 The installation shown in Figure 17 is used for profile milling of compressor blades and consists of a milling machine, and a hydraulically powered robot which loads and unloads the milling machine. The robot is track-mounted in order to service three further profile milling machines planned at the installation. The robot is enclosed by a 2 m-high perimeter fence with access by a door interlocked with a trapped key exchange interlocking system. The milling machine is separately guarded and shares a common boundary with the robot enclosure at one point.

14 The robot loads and unloads the milling machine through an opening in the common boundary fence, and this opening is provided with an interlocked rise and fall powered guard.

## Installation operation

15 A horizontal circular loading table, with one half outside the perimeter fence, is manually loaded with blank components. The loading table is then power indexed through 180° to present components to the robot on the inside of the enclosure.

16 The robot picks up blank components and moves along the track, rotates about its central axis and loads the milling machine through the opening in the common boundary perimeter fence. Unloading at the milling machine is the reverse of this loading procedure.

17 The robot is programmed by means of a teach pendant inside the enclosure.

## Safety arrangements for the operator

*Hazards: Injury by collision with the robot or by trapping between the robot and a fixed object; and ejection of components*

18 The following safeguards are provided:

- 2 m-high perimeter fence forming the robot enclosure and robustly constructed to contain ejected components; and
- the access door into the robot enclosure is mechanically locked shut by means of a trapped key interlock system.

*Hazard: Injury by trapping between the loading table and the opening in the perimeter fence*

19 The following safeguards are provided:

- hinged panels are provided at the loading table which are interlocked with positive operating position detection switches at each hinge. Movement of a hinged panel removes, by electrical control means, the power to the loading table and the robot hydraulic control valves;
- the loading table has conventional hardwired control interlocking; and
- robot interlocking is achieved by using the method set out in paragraphs 181-182 of the main text.

## Safety arrangements for the setter/ programmer and maintenance personnel

*Hazard: Injury by collision with robot or trapping between robot and fixed object*

20 The following safeguards are provided:

- to gain access to the robot enclosure, a key is required to unlock the access door. This key is trapped when the robot is an automatic mode enabled state. Turning the trapped key to the automatic mode inhibited state allows it to be released. This then removes electrical power from the robot hydraulic control valves such that they are spring-returned to block the inlet and outlet ports of the actuators and bypass hydraulic pressure to tank. The released key is used to unlock the access door. When the door is unlocked, a further key is released which is retained by the programmer so that no one else can lock the door closed and restart the robot (see paragraph 188 in the main text);
- robot movement in the 'automatic mode inhibited state' can only be achieved by using the teach pendant. This restores power for programming to the robot but the resultant movement is restricted to a slow speed. The pendant has hold-to-run teach controls, enabling device and a hardwired emergency stop button. Restriction to slow speed movement is achieved and maintained by the robot controller;
- when the teach pendant is used, electrically-operated interlocked shot-bolts are

activated to prevent robot movement along its track. To achieve movement along the track for teaching purposes, the programmer needs to be outside the enclosure with the access doors locked shut and the robot positioned from the main control panel;

- safe systems of work ensure that the programmer is always being observed by a second person outside the enclosure during teaching; and
- hardwired emergency stop buttons are provided inside and outside the enclosure. Actuation causes hydraulic power to be removed by dumping the pressure line and switching off the pump. (The dump valve is held in the non-dump condition electrically during normal running.)

### *Hazard: Injury by trapping between the robot arm and the milling machine*

21 The following safeguards are provided:

- to gain access to the machining area of the milling machine, the rise and fall guard between it and the robot enclosure needs to be in the fully down position. Special interlocking arrangements ensure that the rise and fall guard remains in this position when the milling machine guards are open; and
- the rise and fall guard is strengthened to withstand uncovenanted impact from the robot arm.

### *Hazard: Injury by entanglement or trapping by the milling machine*

22 The following safeguards are provided:

- manual controls are provided at the milling machine inside the enclosure for operating the machine when manual mode is engaged at the main control panel; and
- all dangerous parts of the milling machine are guarded. The cutters are provided with hinged interlocked guards which prevent access to the machining area when the cutters are rotating and when the robot is loading or unloading a component.

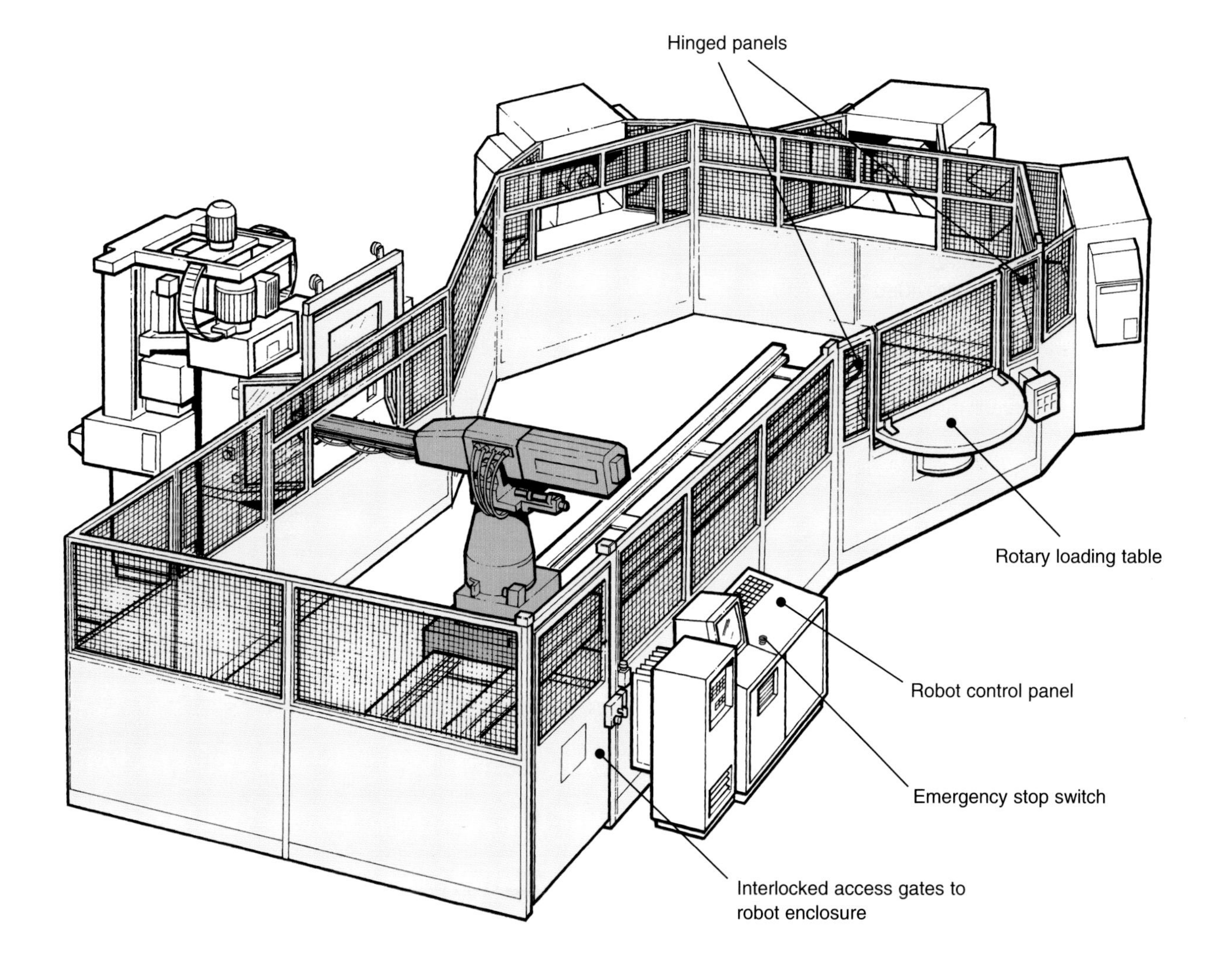


Figure 17 Case study 2: Machining installation

# Case study 3: Robot servicing injection moulding machine and ancillary equipment

## General description of installation

23 A track-mounted, electrically powered robot is used to unload components from a large, plastic injection moulding machine and transfer them through two ancillary machines (a trimming press and an insert machine) before placing them onto a powered cooling conveyor. An outline drawing of the installation is shown in Figure 18.

24 The robot, ancillary machines and the loading end of the cooling conveyor are enclosed by a 2 m-high perimeter fence, which abuts the injection moulding machine. Guarding standards for this machine are those required for manually-operated machines. Programming and maintenance personnel gain access into the enclosure through a door interlocked with a trapped key exchange system. Inside the enclosure is a further trapped key system which allows power to be restored to the robot or the trimming press or the insert machine. Power can only be restored to one machine at any one time.

## Installation operation

25 When the injection moulding machine, powered access guard and tools have opened, the robot removes the moulded component and moves along the track to place it into the trimming press. After trimming, the component is transferred to a second machine for the addition of inserts before being placed aside on the cooling conveyor. This conveyor moves one station each cycle to eventually discharge components outside the enclosure.

26 Trimming and insert addition operations are carried out by the robot during the moulding cycle time. Both these machines are pneumatically powered.

27 Programming is accomplished inside the enclosure by means of a hand-held teach pendant.

28 Access into the enclosure is needed for robot programming purposes, and for setting the injection moulding machine and the two ancillary machines. Access is also required by maintenance personnel.

## Identification of hazards and assessment of risk

29 Hazards were identified to ascertain which parts of the installation might be potential sources of injury. When that had been completed, a simple risk assessment of possible injuries and severity of injury was applied and the results used to decide the level of protection required for each hazard.

30 In normal production the whole installation contains many hazards and the risk assessment was that the foreseeability of injury is high. This initial assessment showed that a high degree of integrity in the perimeter guarding and interlocking arrangements was required.

31 Subsequent identification of other hazards within the robot cell when access is necessary for setting, programming etc influenced the final choice of the method of interlocking the perimeter fence and decided what further protection was needed at specific locations for the setters and programmers.

32 Within the enclosure, access is needed to all of the ancillary machines either for setting, scrap removal, or insert loading. The injection moulding machine presents a high risk of injury and so the existing high-risk interlocking arrangements were retained. On the trimming and insert loading presses, however, the risk assessment indicated a lower chance of injury than if the machines were on a production line and were being hand fed and operated. The standard of protection was chosen as suitable for low risk.

33 The following paragraphs identify each of the main hazards and indicate the detailed safeguards which were selected following hazard identification and risk analysis.

## Safeguarding arrangements for operating personnel

*Hazards: Injury by collision with robot or by trapping between the robot and a fixed object; and injury by entrapment with associated machinery*

34 The following safeguards are provided:

- 2 m-high perimeter fence; and
- to prevent inadvertent access to the robot working area, the access door is mechanically locked shut and the key issued according to a safe system of work when access is required.

## Safeguarding arrangements for the programmer/ maintenance personnel

### *Hazard: Injury by collision with robot or by trapping between the robot and a fixed object*

35 The following safeguards are provided:

- situated at the main control panel outside the enclosure is a trapped key operated switch which, when in the 'automatic mode enabled' state, allows the total installation to operate and produce moulded components. Following the request for access into the safeguarded enclosure, rotation of the key removes, by control means, the power to the axis servo drives. Automatic mode is then inhibited and the key can be removed. The released key can then be used to unlock a mechanical lock on the access door; this key is then trapped in the lock;
- as the access door is unlocked, a further key is released from the key exchange system (see paragraph 188 in the main text). This key is used to restore power for programming but the result is that the robot is limited to slow speed. This restriction in speed is achieved by the robot motion controller. This arrangement is described in paragraph 70 of the main text. While in the teach mode, movement of the robot along its track is not possible;
- the teach pendant has an enabling device and hold-to-run controls;
- hardwired emergency stop controls are provided on the teach pendant, at appropriate points inside the enclosure and on the main control console outside the enclosure;
- when setting or adjusting the injection moulding machine, a safe system of work is used to ensure that the robot is electrically isolated to prevent any movement.

### *Hazard: Injury by entanglement with, or trapping by, ancillary machine*

36 The following safeguards are provided:

- when the automatic mode inhibited state is engaged, separate, independent, hardwired interlocking arrangements at the two ancillary machines cause the pneumatic control valves to block the incoming air supply and dump any stored pneumatic energy;
- the key which is released when the access door is unlocked can also be used to reinstate control power to either the trimming press or the insert machine by choosing the appropriate keyed switch inside the enclosure. Manual control is then possible at the control box situated alongside each machine;
- a risk assessment of the trimming press indicated that injury is foreseeable when operated inside the enclosure. Therefore a pressure-sensitive mat surrounds the machine and the control panel located outside its perimeter. Single control system interlocking standards are appropriate in this instance; and
- emergency stops are provided at each machine.

### *Hazard: Injury by trapping on the injection moulding machine*

37 The following safeguards are provided:

- the injection moulding machine retains in full all the high-risk interlocking safeguards required at a stand-alone installation; and
- when programming the robot, a safe system of work is used to ensure that the injection moulding machine is electrically isolated to prevent any movement.

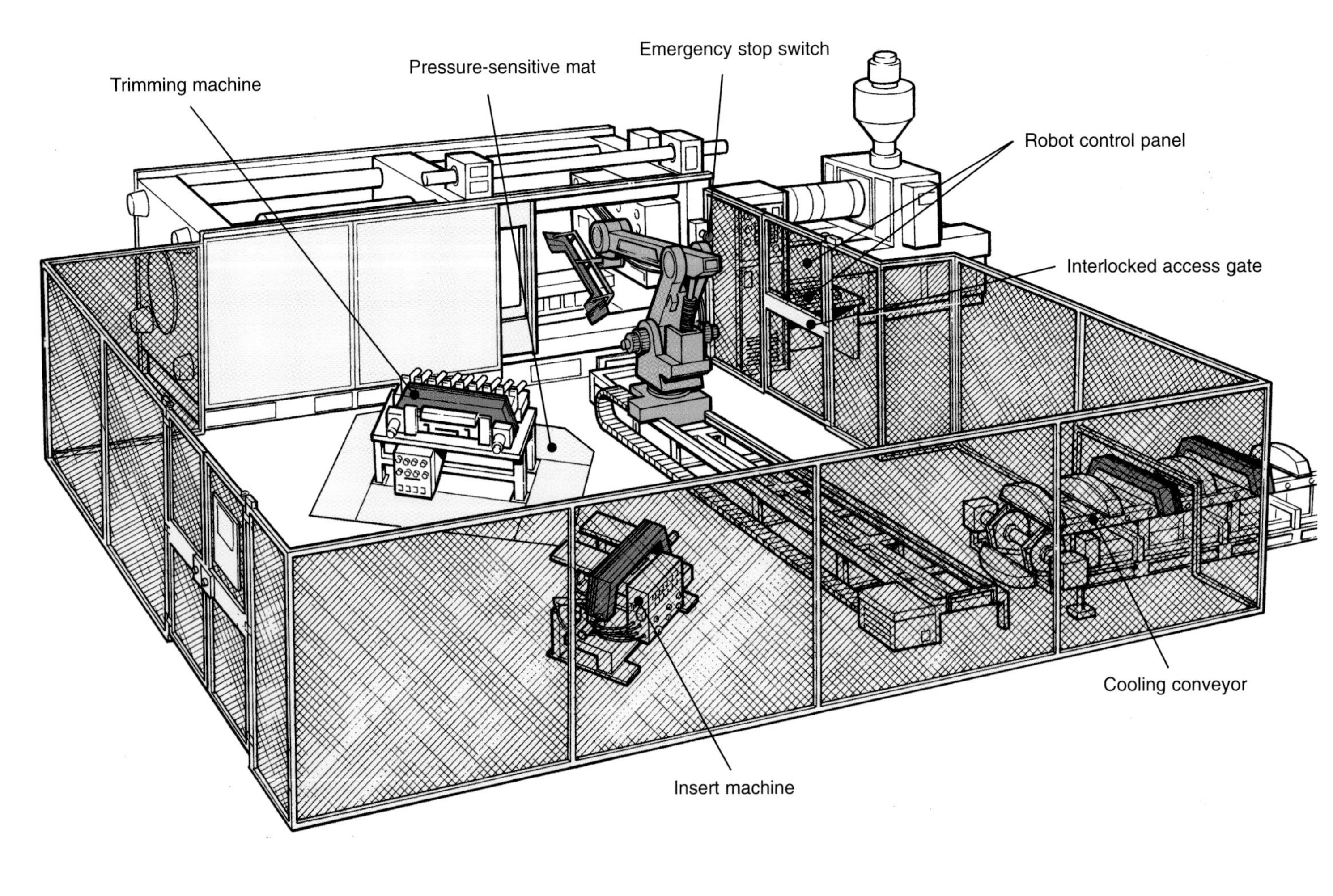


Figure 18 Case study 3: Injection moulding machine and ancillary equipment

# Case study 4: Robot welding installation

## General description of installation

38 An outline drawing of the installation is shown in Figure 19. It shows an electrically driven robot with five axes of movement, together with a MIG welding set. The robot is bolted to the floor and rotates about a central axis to weld at any one of the three fixed position welding stations.

39 The robot, robot control cabinet, welding set and welding stations are enclosed by a 2 m-high sheet-steel perimeter fence. Access to load and unload components at the welding stations is by interlocked manually operated counterbalanced rise and fall guards. Access for programming and maintenance purposes is by a door interlocked with a trapped key interlock system.

## Installation operation

40 When a welding cycle is in progress at a particular station, the vertical guard there is kept closed by two shot-bolt devices operated by a spring (only one shown for clarity). The component clamps are held on by air pressure.

41 At the end of the welding cycle, all air pressure to the clamps is dumped to exhaust and air pressure applied to the shot-bolt devices to release them. The counterbalanced vertical guard can then be raised manually to remove the completed component. To achieve this, the clamps have to be manually withdrawn.

42 Components for welding can then be loaded into the welding fixture and the clamps positioned manually on to the components (ie without air pressure). A two-hand control is then operated to pressurise the pneumatic clamps so they hold components in position. The location of the component can be visually checked prior to the welding cycle.

43 To start a welding cycle at a particular station the vertical guard needs to be manually closed and a cycle initiation button activated to dump to exhaust air pressure in the shot-bolt devices. The position of each shot-bolt is monitored to ensure that the guard is closed.

44 Programming is done with a hand-held teach pendant, usually from outside the enclosure. Access for teaching inside the enclosure is by the interlocked door. Access to the enclosure is required to change welding wire reels, but gas cylinders are situated outside the perimeter fence. All robot movement and welding controls are situated on the teach pendant.

45 Nozzle cleaning equipment is provided within the enclosure and the robot needs to be positioned using hand controls when the equipment is used.

## Safeguarding arrangements for the operator

### *Hazard: Injury by collision with the robot or by trapping between the robot and a fixed object*

46 The following safeguards are provided:

- 2 m-high perimeter fence;
- interlocked rise and fall vertical guards provide access for component loading/unloading, and monitored shot-bolt devices (two per station) keep the guards locked shut when the robot is welding at a particular station. Two independent programmable electronic systems are used. In this case, the robot controller acts as system 1 and an independent programmable logic controller as system 2. Each system monitors the operation of one of the shot-bolts at each station. For example, at one station system 1 is informed that the guard is open when monitored shot-bolt 1 is activated to unlock the rise and fall guard. It then ensures, through software control, that the robot will not enter the zone of that welding station. If any fault causes the robot arm to enter the welding station, the robot controller will effect a controlled shutdown of the robot. Programmable electronics system 2 is informed of guard position by monitoring shot-bolt 2 and also receives robot arm position signals. After a specified short delay to allow system 1 to operate, this second system will de-energise the motive power to each robot axis if the robot enters the welding station. Thus system 2 protects against any fault in the robot control system. This function is repeated at the other two stations. The principle of this safeguarding method is given in paragraphs 196-197 and Figure 14 in the main text; and
- trapped key interlock system access door provides access for welding wire replenishment, cleaning etc (see 'Safety arrangements for the programmer/maintenance personnel' on page 42 for details).

### *Hazard: Trapping by pneumatically operated clamps*

47 The following safeguards are provided:

- air pressure at the clamps is dumped before the vertical screenguard is raised; and
- the clamps are applied manually before air pressure is introduced by means of a two-hand control device.

## Safeguarding arrangements for the programmer/ maintenance personnel

### *Hazard: Injury by collision with robot or by trapping by robot*

48 The following safeguards are provided:

- access into the robot enclosure is obtained via a trapped key interlock system. Following a request for access into the safeguarded enclosure, rotation of the key removes, by control means, the power to the axis servo drives. The key can now be removed and used to unlock the access door (see paragraph 188 of the main text). A further key is released when the door is unlocked and this is used to restore power for programming with the result that the robot is limited to slow speed (see paragraph 70 of the main text). When the power for programming has been restored, a key is released which is retained by the programmer;
- the teach pendant has an enabling device and hold-to-run control. Both hands are needed during the teach mode;
- the robot will only operate at the reduced speed, except that welding deposition (which is undertaken at speeds less than 0.25 m/s) can still be undertaken at normal operating speed;
- automatic mode is not attainable; and
- hardwired emergency stops are provided outside the enclosure at each welding station and also on the teach pendant.

### *Hazard: Trap by clamp*

49 The following safeguards are provided:

- pneumatically powered movement of the clamps can only be achieved from outside the enclosure; and
- a safe system of work ensures that the air supply is isolated and locked off when any person is inside the enclosure.

### *Hazard: Arc glare*

50 The following safeguards are provided:

- software interlocking should be used to inhibit welding in teach mode during the early stages of programming; and
- close observation of the welding function during the later stages of programming will necessitate the programmer using personal protective equipment such as a welding helmet.

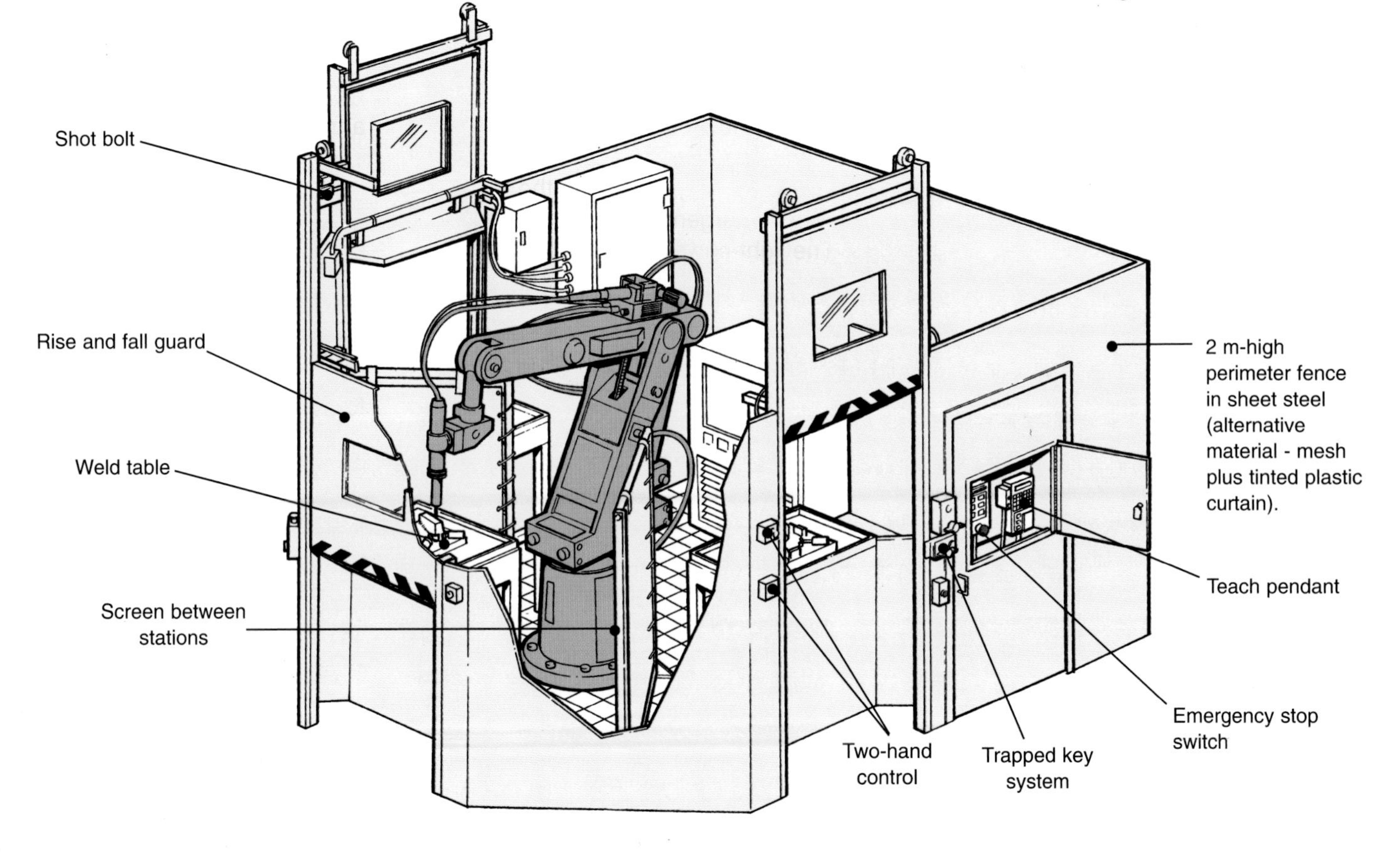


Figure 19 Case study 4: Robot welding installation

# Case study 5: Robot in education establishment

## General description of installation

51 An electrically powered industrial robot is installed in the engineering laboratory of an educational establishment where it is used:

- as a research tool for academics and postgraduate students,
- for demonstration; and
- in a variety of undergraduate laboratory experiments.

52 This wide range of use of the robot created important requirements in the application of safeguards. These were:

- ease of access to the robot working space because of the quantity and diversity of ancillary equipment used with the robot;
- a good view of robot movement from the perimeter of the safeguarded space; and
- high integrity due to the frequent presence of large groups of people, mainly students.

53 The installation is fitted with a photoelectric light curtain and with an interlocked access gate for controlled entry into the safeguarded area.

## Installation operation

54 The robot is operated from a main control panel, keyboard and VDU situated outside the safeguarded area and adjacent to the access gate. Only when this gate is closed and locked and the photo-electric curtain activated can the robot be powered-up. A password/code also needs to be keyed before access is achieved to the operating program and movement of the robot initiated.

55 Semi-automatic, manual and programming operations are carried out from the main control panel. Automatic operation, although possible, is used infrequently. When close approach during programming is necessary, a hand-held teach pendant is used inside the safeguarded area.

## Safeguarding arrangements when the robot is operated from main control panel

*Hazard: Injury by collision with the robot or by trapping between the robot and a fixed object*

56 The following safeguards are provided:

- a high-integrity, self-checking, photoelectric curtain is mounted around the perimeter of the installation - this has two channel outputs hardwired directly into two final switching devices in the robot emergency stop circuit. The light curtain is installed in accordance with the principles set out in HSE Guidance Note HSG180.[6] To minimise the possibility of inadvertent interruption of the light curtain a line, indicating the limit of approach for students/other observers, is marked on the floor and walls;
- the access gate adjacent to the control panel is mechanically locked shut by means of a trapped key interlock system (see programming safeguards below for explanation); and
- a safe system of working is followed which includes:
  - operation of a keyed switch in the main control panel to apply power to the robot (the key is retained by a responsible person when the installation is not in use); and
  - the use of the password (or code) which needs to be keyed in before access to the robot-operating program can be achieved.

## Safeguarding arrangements during programming

*Hazard: Injury by collision with the robot or by trapping between the robot and a fixed object*

57 The following safeguards are provided:

- the robot may be programmed from the main control panel when the safeguards already described will apply. If it is necessary to approach the robot closely during programming, the following safeguards will also apply;
- controlled access to the robot working space is achieved via a trapped key interlock system. Following a request for access into the safeguarded enclosure, rotation of the trapped key unlocks the gate and releases a second key which is taken by the programmer into the safeguarded area and is used to activate the teach pendant system (see paragraph 188 in the main text). The robot may then only be operated from the teach pendant and movement is restricted to slow speed. An external safety system monitors this slow speed (see paragraphs 193-195 and Figure 13 in the main text);
- the teach pendant is fitted with:
  - an enabling device;
  - hold-to-run controls; and
  - a hardwired emergency stop;

- the swept area of the robot is painted on the floor; and
- a safe system of work is provided.

58 Note that in this case study ejection was not considered to be a significant hazard. Where ejection is a hazard, physical barriers in addition to the light curtain, or as an alternative, should be provided.

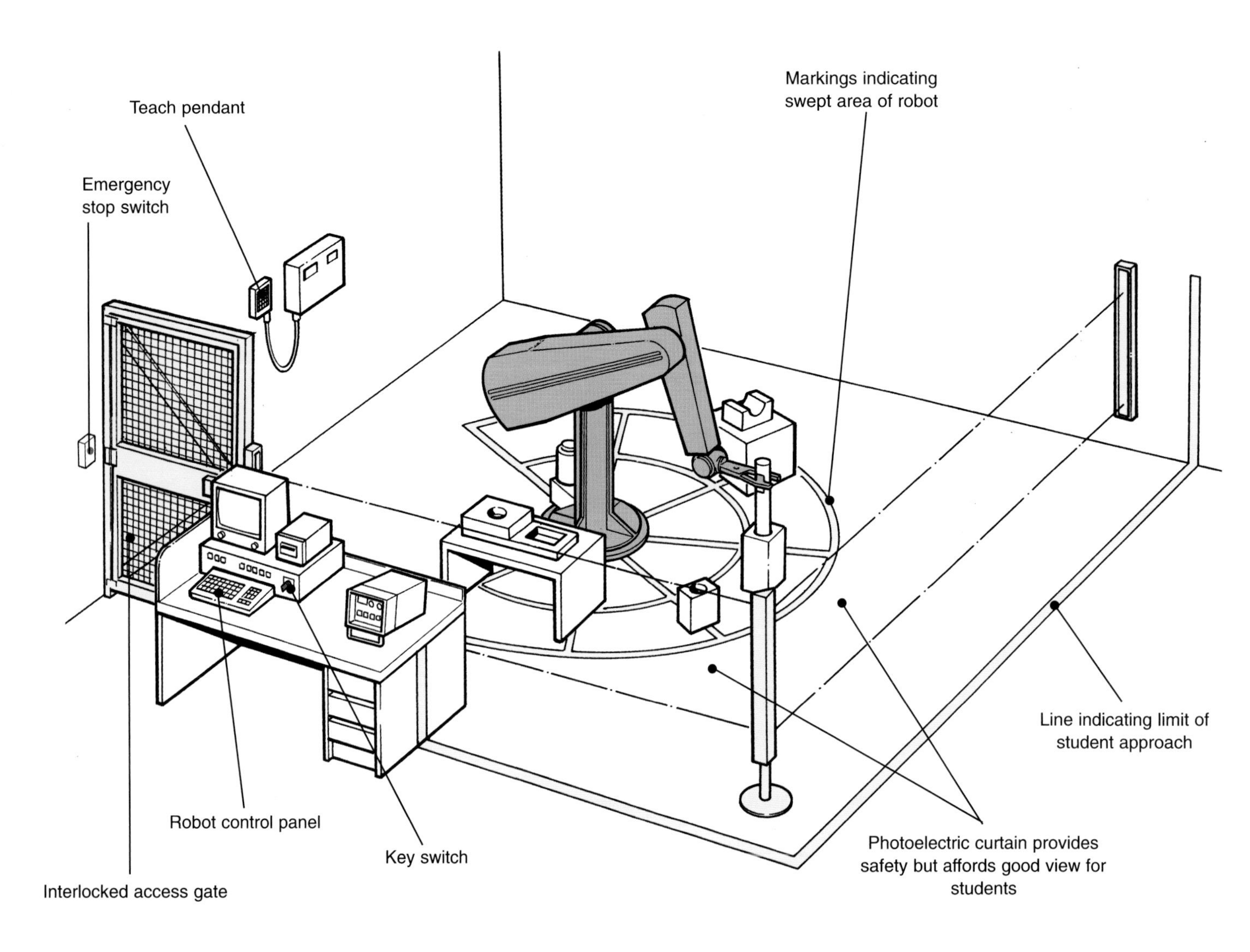


Figure 20 Case study 5: Robot in education establishment

# Case study 6: Robot water jetting

## General description of installation

59 A hydraulically powered robot is used to clean various large, intricate and high-value components using a high-pressure water jet operating at about 500 bar. The robot and component to be cleaned are contained in a totally enclosed chamber with all power supplies and control systems situated outside the chamber.

60 Access to load/unload components and for programming and maintenance is by double doors fitted with a trapped key interlock system.

## Installation operation

61 The component to be cleaned is mounted on a fixture which is accurately positioned in relation to the robot. Only when the double doors are closed and locked can the robot and water pump be started by means of key operated switches.

62 The robot program comprises 24 part programs, each corresponding to an area to be cleaned. If a particular area is not cleaned to the required quality standard, it can be recleaned by selecting the particular part program.

63 Programming is achieved with a hand-held pendant inside the enclosure.

64 To prevent damage to expensive components by the fast-moving robot, the gripper end of the robot arm is designed to collapse on impact, under spring pressure, and actuate micro-switches to stop the robot. No protection to personnel is offered by this arrangement.

## Safeguarding arrangements for the operator

*Hazard: Injury by collision with the robot or by trapping between the robot and a fixed object*

65 The following safeguards are provided:

- a totally enclosed chamber;
- the access doors to the enclosed chamber are mechanically locked shut by means of a trapped key interlocked system. Two keys are needed to unlock the doors; one key is associated with the robot and the other with water pump power switching. Turning the trapped key associated with the robot puts, via the robot motion controller, the axis motors into servo-hold. The key associated with the water pump can only be released when the power to the pump has been isolated. Opening the access door releases a third trapped key and this is kept by the operator to prevent the doors being locked closed while working inside the robot chamber.

*Hazard: Injury by high-pressure water jet*

66 The following safeguards are provided:

- a totally enclosed chamber; and
- before the access door can be opened, the water pump power switching device has to be switched off by turning the trapped key, and the released key transferred to the access door interlock. Both keys, the robot activation device key and the water pump key, have to be transferred to the access door interlock before the doors can be opened. The hardwired power interlocking arrangements detailed above are appropriate to this hazard.

*Hazard: Hearing impairment by noise generated by high-pressure water jet*

67 The following safeguard is provided: a totally enclosed chamber reduces noise to an acceptable level.

## Safeguarding arrangements for the programmer/ maintenance personnel

*Hazard: Injury by collision with the robot or by trapping between the robot and a fixed object*

68 The following safeguards are provided:

- as previously described, turning the trapped key at the robot activation device to release it puts the axis motors into servo-hold. After gaining access to the enclosure for programming or maintenance purposes, robot movement can only be achieved by means of the teach pendant and only at reduced speed. Activating the teach pendant system allows the robot controls on the teach pendant to be operated. An external safety system monitors the slow speed movement;
- the teach pendant has an enabling device and a hold to run self-centreing joystick control. Both hands are needed during the teach mode;
- a lock-off type isolator is provided to electrically isolate the robot;

- the working area of the robot is painted on the enclosure floor;
- hardwired emergency stops are provided to remove all electrical power;
- blocking valves are installed on the robot to prevent droop when the hydraulic power is switched off; and
- safe systems of work are provided in writing.

## *Hazard: Injury by high-pressure water jet*

69 The following safeguards are provided:

- the trapped key at the water pump power switching device is needed to open the enclosure access doors; and
- to release the key the pump has to be switched off. This hardwired power interlocking arrangement is appropriate for the hazard.

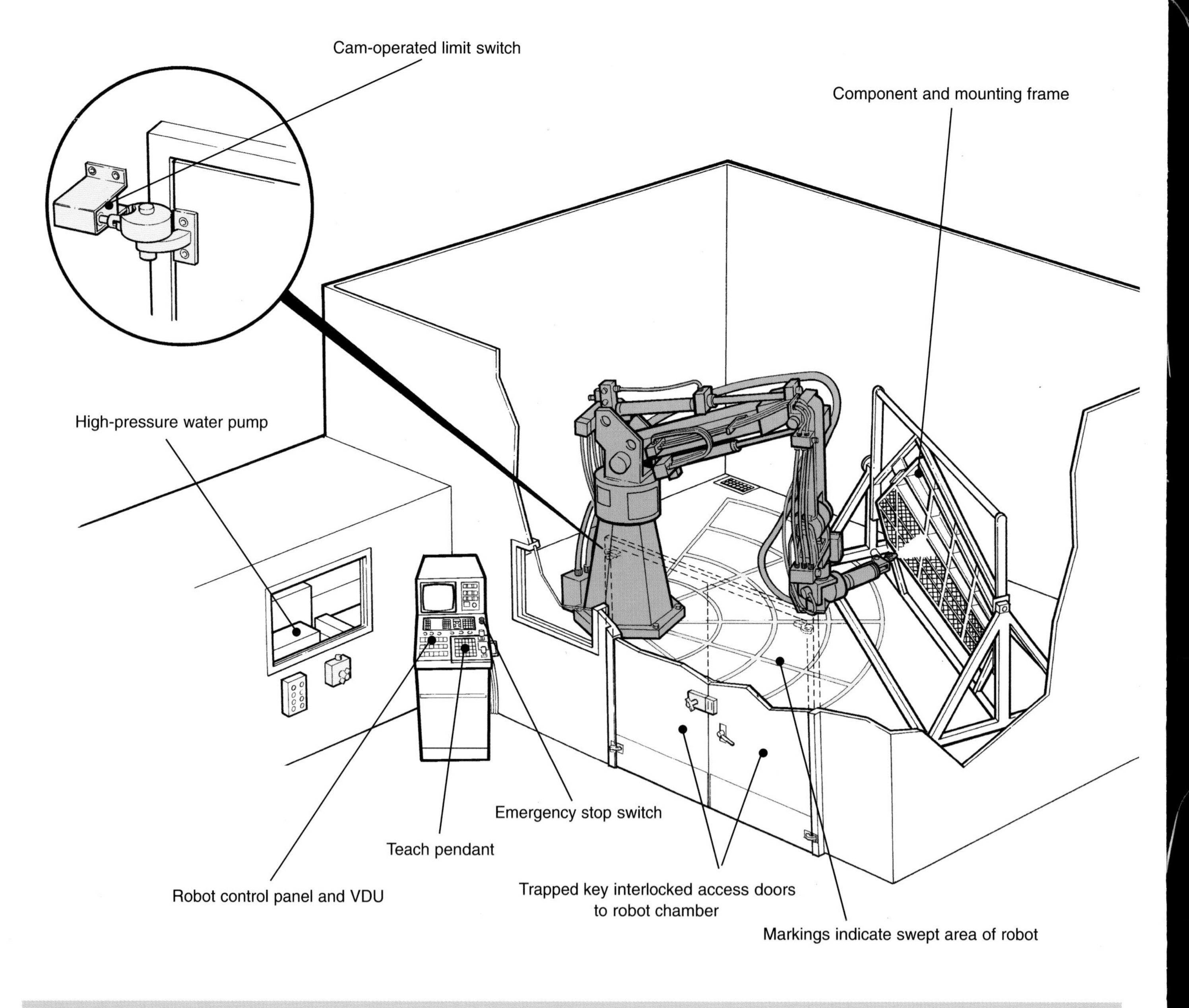


Figure 21 Case study 6: Robot water jetting

# Case study 7: Robot palletising installation

## General description of installation

70 An electrically powered robot is used to palletise 25 kg bags of two different products. The installation comprises two product in-feed conveyors supplying bags to the robot pick-up conveyors ready for stacking onto pallets. A full pallet is made up of eight layers of five bags per layer, each layer having an alternating layering pattern for load stability. Associated machinery includes pallet dispensers, slip sheet applicators, pallet conveyors and a stretch-wrapping machine.

71 The robot and associated machinery is enclosed by a 2 m-high perimeter fence equipped with access points for bag in-feed and loaded pallet output conveyors. The installation is divided into two safety zones separated by 2 m-high fencing with pallet transfer access between zones for loaded pallets. Bag conveying, robot palletising and pallet conveying machinery (full and empty) are contained in Zone A. Pallet conveying and stretch-wrapping machinery are contained in Zone B. The pallet conveyor between Zones are common to both. Access to the enclosure for operating and maintenance personnel is through interlocked gates in the fence.

## Installation operation

72 Full bags from product line 1 are conveyed to the robot pick-up point and are picked up by the robot and placed onto an empty waiting pallet located at the loading position for product line 1. The robot will continue to stack bags according to the required loading pattern until the pallet is full. Product line 2 operates in a similar fashion. Full pallets are then conveyed either into Zone B to the stretch-wrapper turntable or via a shuttle conveyor to a fork-lift truck (FLT) pick-up point in Zone A. Stretch-wrapped pallets are conveyed to a separate FLT pick-up point in Zone B. Full (unwrapped) pallets from the warehouse can be reloaded by FLT via the shuttle conveyor into Zone A for stretch-wrapping.

## Safeguarding arrangements for the operator/ programmer/ maintenance personnel

*Hazards: Injury by collision with the robot, or by trapping between the robot and a fixed object; and injury from contact with associated machinery*

73 The following safeguards provide protection from the robot palletiser and associated machinery.

### *Safeguarding fence*

- The robot and associated machinery are enclosed within a 2 m-high perimeter fence. The fenced area is divided into two safety zones (A and B) by a 2 m-high fence. Pallets are transferred between the two zones through an access opening in the fence which is protected by a light curtain. This light curtain is muted during pallet transfer.
- Access to the fenced area is possible through gates in the fence (which are hardwired interlocked) and the FLT pick-up points (which are protected by light curtains). In either case, the safeguards detailed in 'Safeguarding zones' below will prevail. Each light curtain is muted when the presence of a loaded pallet is detected at the pick-up point. If a light curtain is interrupted other than by a pallet (ie attempted entry by a person), all machinery in the zone will be stopped. An actuated light curtain needs to be reset before power can be restored to machinery within the respective zone.
- Fence access gates are also equipped with a mechanical latch to prevent them opening accidentally or their closing unintentionally while someone is inside the fenced area.

### *Safeguarding zones*

- Zone A contains the robot palletiser and associated machinery. All access gates and light curtains to Zone A are interlocked with the robot palletiser and associated machinery through PLC software and through hardwired, electromechanical equipment (see paragraph 181 and Figure 7 in the main text). The stretch-wrapper is not interlocked with Zone A machinery except when an emergency stop button is actuated.
- Zone B contains the stretch wrapping machine and associated machinery. All access gates and light curtains to Zone B are interlocked with the stretch-wrapper and associated machinery through PLC software and through hardwired, electromechanical equipment. The robot is not interlocked with Zone B machinery except when an emergency stop button is activated.

### *Operational safeguards*

- Electrical power cannot be made available to the controls and equipment within the zones until all the following conditions have been met:

- the contactors on the motor control centre are closed ('ON');
- individual isolators of associated machines are closed ('ON'); and
- the robot control captive keyswitch is in the 'ON' position.

- The robot and associated machinery can only be started from the touch screen on the operator/robot control interface panel and palletising can begin only when:

- the robot keyswitch is set to 'ON';
- the robot status switch is set to automatic;
- all gates in the fence are closed;
- all light curtains have been reset; and
- all emergency stop buttons have been released.

- In normal operation, a controlled shutdown of the robot through the robot controller is desirable before entry to the robot enclosure. A controlled shutdown allows the robot to complete its current cycle and stop in its home position via software control and cuts off power electromechanically to its servo-drive units.
- An uncontrolled shutdown caused by opening a gate in the fence or a person interrupting a light curtain, will stop all machinery in the relevant zone (including the robot in Zone A and stretch-wrapper in Zone B). The robot will shut down immediately in mid-cycle via software control and power to its servo-drive units will be electromechanically cut off. An uncontrolled shutdown caused by operating an emergency stop button will immediately stop all machinery in both zones. Restarting the robot after an uncontrolled shutdown can only be achieved via the reset procedure at the touch screen on the operator/machine interface panel.

## *Teaching*

- Teaching the robot can only be achieved with the robot status switch in manual control. When under manual control, the robot will only move at slow speed. Teaching is carried out using a teach pendant and when the teach pendant is being used, the robot cannot be controlled from the touch screen at the operator/robot interface panel. Most, if not all, teaching can be done from outside the safeguarded enclosure, but close observation of the robot from inside the enclosure may be necessary during teaching or for maintenance. This requires a safe system of work which demands two people to be present, one working within the enclosure and the other observing operations from outside.

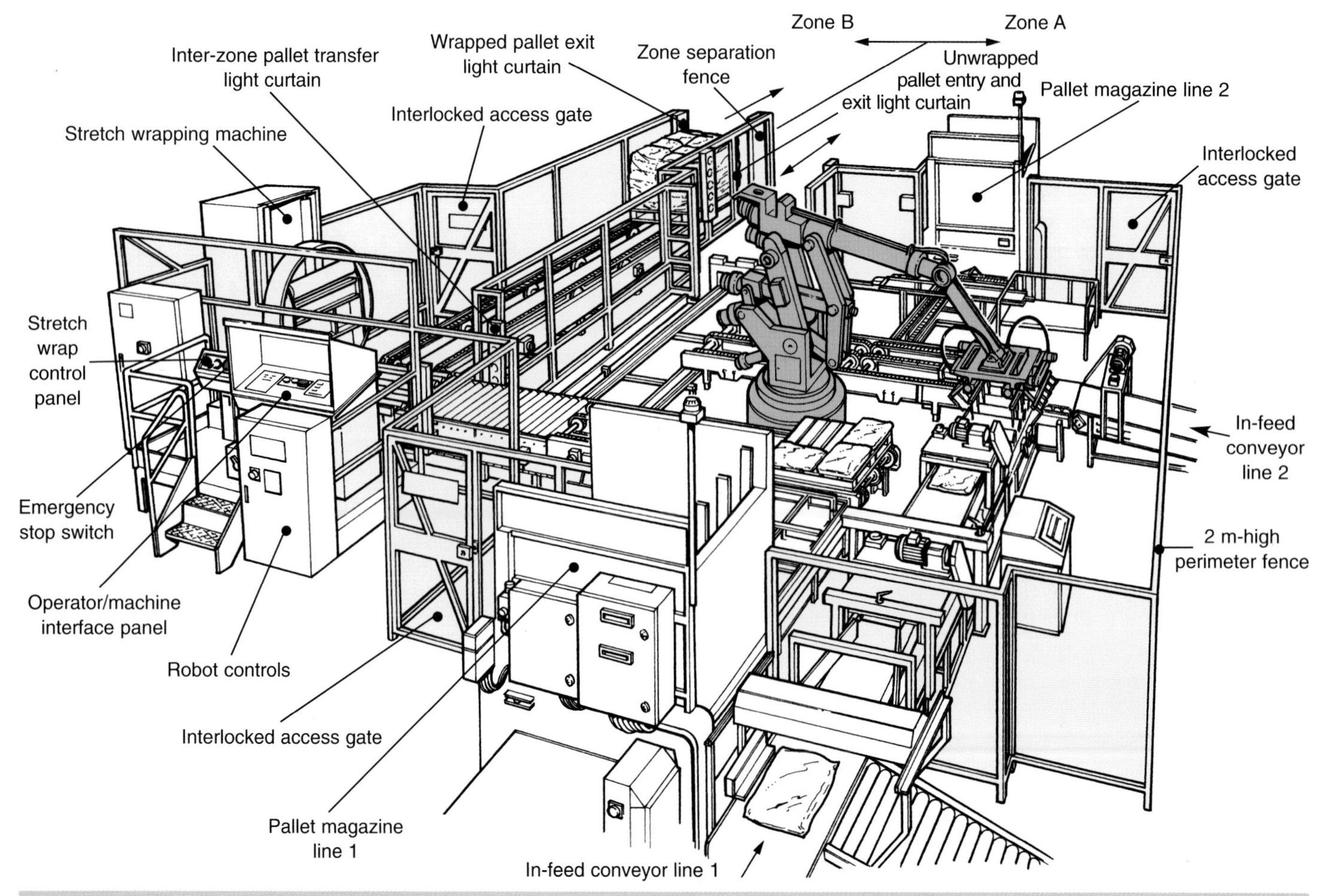


Figure 22 Case study 7: Robot palletising installation

# References


1 *The Health and Safety at Work etc Act 1974* The Stationery Office 1974 ISBN 0 10 543774 3

2 *Safe use of work equipment. Provision and Use of Work Equipment Regulations 1998. Approved Code of Practice and Guidance* L22 HSE Books 1998 ISBN 0 7176 1626 6

3 *Supply of Machinery (Safety) (Amendment) Regulations 1994* SI 1994/2063 Stationery Office 1994 ISBN 0 11 045063 9

4 *Safe use of lifting equipment: Lifting Operations and Lifting Equipment Regulations 1998: Approved Code of Practice and Guidance* L113 HSE Books 1998 ISBN 0 7176 1628 2

5 *Management of health and safety at work. Management of Health and Safety at Work Regulations 1999. Approved Code of Practice* L21 (Second edition) HSE Books 2000 ISBN 0 7176 2488 9

6 *Five steps to risk assessment* INDG163(rev1) HSE Books 1998 Single copies free, multiple copies in priced packs ISBN 0 7176 1565 0

7 *Application of electro-sensitive protective equipment using light curtains and light beam devices to machinery* HSG180 HSE Books 1999 ISBN 0 7176 1550 2

8 *The Health and Safety (Enforcing Authority) Regulations 1998* SI 1998/494 Stationery Office 1998 ISBN 0 11 065642 3

9 *The Electrical Equipment (Safety) Regulations 1994* SI 1994/3260 Stationery Office 1994 ISBN 0 11 043917 1


# Relevant standards

These standards are arranged in numerical order.


BS EN 292: *Safety of machinery - Basic concepts, general principles for design* Part 1: 1991 *Basic terminology, methodology* Part 2: 1991/Amd A1: 1995 *Technical principles and specifications*

BS EN: 294: *Safety of machinery - Safety distances to prevent danger zones being reached by the upper limbs*

BS EN 349: *Safety of machinery - Minimum gaps to avoid crushing of parts of the human body*

BS EN 418: *Safety of machinery - Emergency stop equipment, functional aspects - principles for design*

BS EN 574: *Safety of machinery - Two-hand control devices - Functional aspects - Principles for design*

EN 775: 1992 - *Manipulating industrial robots - Recommendations for safety*

BS EN 953: 1998 - *Safety of machinery - Guards - General requirements for the design and construction of fixed and movable guards*

BS EN 954: *Safety of machinery - Safety-related parts of control systems* Part 1: *General principles for design*

BS EN 999: 1998 - *Safety of machinery - The positioning of protective equipment in respect of approach speeds of parts of the human body*

BS EN 1050: *Safety of machinery - Principles for risk assessment*

BS EN 1088: *Safety of machinery - Interlocking devices associated with guards - Principles for selection*

BS EN 1760: *Safety of machinery - Pressure-sensitive protective devices* Part 1: *General principles for the design and testing of pressure-sensitive mats and pressure-sensitive floors*

BS 5304: 1988 - *Code of practice for safety of machinery* (now obsolescent but will remain available to provide guidance on guarding older machines)

BS EN 60204: 1998 - *Safety of machinery - Electrical equipment of machines* Part 1: *Specifications for general requirements*

BS EN 61496: 1998 - *Safety of machinery - Electro-sensitive protective equipment* Part 1: *Specification for general requirements and tests*

IEC 61496: 1997 - *Safety of machinery - Electro-sensitive protective equipment* Part 2: *Particular requirements for equipment using opto-electronic protective devices*

IEC 61508: *Functional safety of electrical/electronic/programmable electronic safety-related systems*


# Further reading


Product standards: Machinery - Guidance notes on UK Regulations. Department of Trade and Industry. Tel: 020 7944 4888 for latest version


While every effort has been made to ensure the accuracy of the references listed in this publication, their future availability cannot be guaranteed.

# Standards

These are available from BSI Customer Services, 389 Chiswick High Road, London W4 4AL. Tel: 020 8996 9001 Fax: 020 8996 7001.

# Stationery Office publications

Stationery Office (formerly HMSO) publications are available from The Publications Centre, PO Box 276, London SW8 5DT. Tel: 0870 600 5522 Fax: 0870 600 5533. They are also available from bookshops.

# HSE publications

See back page for details.

# Glossary

## Associated machinery

Machines working in conjunction with the robot within the robot enclosure or interfacing with it, eg input and output conveyors, presses fed by the robot.

## Close approach

The requirement for a person to be in close proximity to the robot when the robot is under power. This may be necessary in the final stages of teaching and during maintenance. Close approach should always be undertaken within an established safe system of work.

## Peripheral devices

In this document, peripheral devices are those which work in synchronisation with the robot and which have their motion managed by the robot motion controller, eg turntables or manipulators.

## Reliability

Reliability (of a machine) is the ability of a machine or its components or equipment to perform a required function under specified conditions and for a given period of time without failing. In this document it refers specifically to failures of hardware which could directly cause a dangerous situation, or allow a dangerous situation to develop. The term includes the failure of electronic components but does **not** include faults or failures in software.

## Risk assessment

A comprehensive estimation of the probability and the degree of the possible injury or damage to health in a hazardous situation in order to select appropriate safety measures.

## Safety integrity

Safety integrity is the probability of a safety-related system satisfactorily performing the required safety functions under all the stated conditions within a stated period of time. For a robot system, the safety integrity will need to include all causes of failures which can lead to an unsafe state, both in hardware and software, as well as failures due to electrical interference.

It may be possible to measure the failure of a hardware item in terms of its known failure rate, but the safety integrity of a system also depends on factors which cannot be accurately measured in this way, eg software and components incorporating software. The major factor to be considered in these aspects of a system is the quality of their manufacture or production.

## Safety-related system

A safety-related system is one that:

- implements the required safety functions which are necessary to achieve, or maintain, a safe state for the robot; and
- is intended to achieve, on its own or with other safety-related systems, the necessary level of safety integrity for the implementation of the required safety functions.

## Teaching

Programming the motion of the robot to carry out its specific task. This is done with the robot control system in teach mode which allows the generation, storage and playback of positional data points while the robot is under slow speed control. Teaching is usually carried out using a hand-held teach pendant (linked to the robot control system) to move the robot through its required actions.

Printed and published by the Health and Safety Executive C70 11/00

**MAIL ORDER**
HSE priced and free
publications are
available from:
HSE Books
PO Box 1999
Sudbury
Suffolk CO10 2WA
Tel: 01787 881165
Fax: 01787 313995
Website: www.hsebooks.co.uk

**RETAIL**
HSE priced publications
are available from
good booksellers

**HEALTH AND SAFETY ENQUIRIES**
HSE InfoLine
Tel: 08701 545500
or write to:
HSE Information Centre
Broad Lane
Sheffield S3 7HQ
Website: www.hse.gov.uk